AF560460

# Fundamentals of Textile Fibre

# Fundamentals of Textile Fibre

Sandeep Roy

**RANDOM PUBLICATIONS**
NEW DELHI (INDIA)

**Fundamentals of Textile Fibre**

---

ISBN 978-93-5111-868-8
© Reserved

All Rights Reserved. No Part of this book may be reproduced in any manner without written permission.

Published in 2016 in India by

**RANDOM PUBLICATIONS**

4376-A/4B, Gali Murari Lal, Ansari Road
New Delhi-110 002
Phone : +9111-43580356, 011-23289044, 011-43142548
e-mail: sales@randompublications.com,
info@randompublications.com, randomexports@gmail.com

Reprinted. 2021

*Type Setting by* : Friends Media, Delhi-110089
*Digitally Printed at* : Replika Press Pvt. Ltd.

# Preface

Textile fibers have been used to make cloth for several thousand years. First manufactured fiber was produced commercially on 1885 and was produced from fibers of plants and animals. Wool, flax, cotton and silk were commonly used textile fibers. Textile fibers are characterized by the flexibility, fineness and large length in relation to the maximum transverse dimension.

Morphology or physical structure of textile fiber includes the study of the size, shape and structure of textile fiber, by observing the fiber using a microscope and the relationship between these properties. Fiber's morphology influences fabric characteristics and performance and the process that will be used in producing a finished fabric.

Textile fiber has some characteristics which differ between fiber to Textile fiber. Textile fiber can be spun into a yarn or made into a fabric by various methods including weaving, knitting, braiding, felting, and twisting. The essential requirements for fibers to be spun into yarn include a length of at least 5 millimeters, flexibility, cohesiveness, and sufficient strength. Other important properties include elasticity, fineness, uniformity, durability, and luster. Most fabric structures are composed of actual fabric rather than meshes or films. Typically, the fabric is coated and laminated with synthetic materials for increased strength, durability, andenvironmental resistance. Among the most widely used materials are polyesters laminated or coated with polyvinyl chloride (PVC), and woven fiberglass coated with polytetrafluoroethylene(PTFE).

Bleaching of Textile Fibres incorporates the necessary information about all the pretreatments given to various textile fibres like cotton wool silk Rayon Polyester Nylon, Acrylic Acetate jute linen etc. This book will be useful to all the students engineers and professionals related to various disciplines of Textiles Fashion and Garment. This book aims to provide some valuable information about textile fibres. A current broad account of the literature is presented in scientific, yet simple way.

*– Author*

# Contents

# 1

# Introduction

## TEXTILE

We find the availability of diverse sort of textiles in the market. But as far as the consumers are concerned they are ignorant of proper knowledge of identification of fabrics, yarns, finishers. The term textile comprises fibres, yarns and fabrics. Fibre is the basic unit of which a fabric is made.

It may be staple, that is of short length like cotton (1.5 to 7 cm) or wool (2 to 20 cm) or filament, that is, continuous in length, like silk. *Yarns* are continuous lengths formed by twisting together staple fibres or several filaments. Yarns made of filament fibres are generally smooth, glossy and strong.

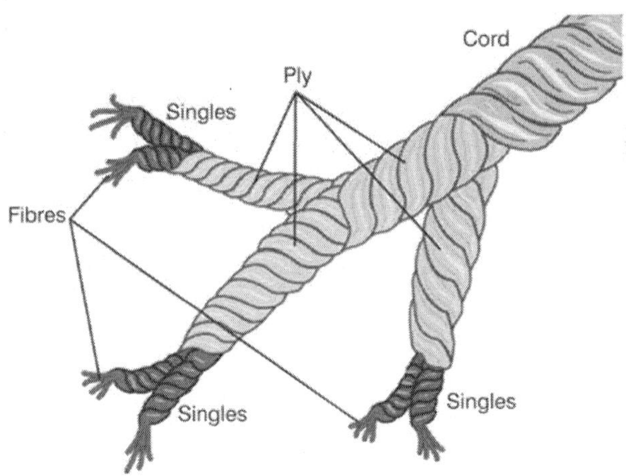


**Fig.** Yarns

Yarns made of staple fibres are more absorbent than filament yarns. The amount of twist provided to a yarn to hold the fibres together influences the smoothness, strength and shrinkage properties of a fabric. High twist yarns, such as those used in crepes, are less lustrous, but have greater strength, elasticity and shrinking properties than low twist yarns. Low twist yarns are used in satin or damask, where a lustrous appearance is desired.

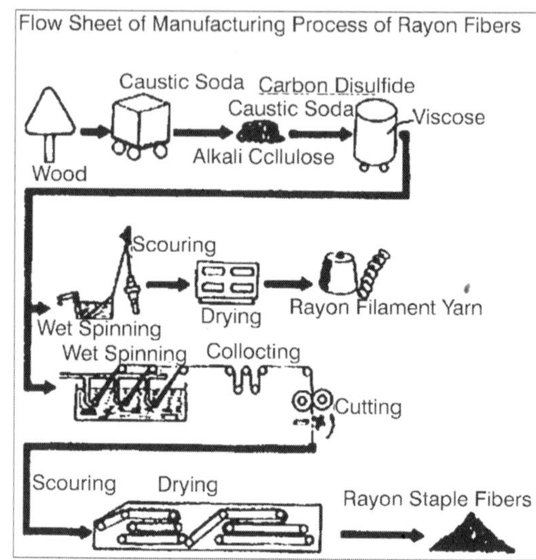


**Fig.** The regenerated fibres

*A fabric* is the product of interlacing or interlocking of yarns or fibres. The interlacing of yarns is called *weaving* and a variety of fabrics are obtained by using different kinds of weaves. Weaves determine many characteristics of a fabric, namely appearance, texture, lustre, durability, elasticity, absorbency, drapability and warmth.

Generally, firm, closely woven fabrics with an equal number of lengthwise and cross-wise yams are strong. Knitting is the second most repeatedly used method of fabric construction.

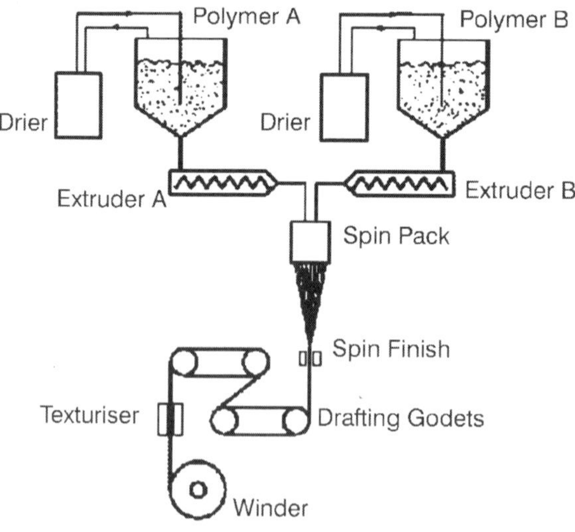


**Fig.** Synthetic fibres

Jersey and tricot are knitted fabrics made by interlocking of a single yarn or a series of single yarns. The more compact the knit construction, the greater will be the durability and shape retention. Until about a century ago, all the fibres used in making fabrics were obtained from natural sources, like plants or animals. Today we have several man made fibres.These are either *regenerated* or *synthetic.* The *regenerated* fibres are made from the same chemical substance, for instance, cellulose from wood pulp which is not in fibre form and is converted to cellulose in fibre form, such as viscose rayon.

*Synthetic* fibres are manufactured from petrochemicals, here the beginning material is totally different from the final product. Polyester, for example, is made from ethylene glycol and terephthalic acid. Among the man-made fibres are also metallic fibres: aluminium, silver, and gold; and mineral fibres: glass and graphite1 Table provides a broad classification of fibres in general. Table. Division of textile fibres in general use for apparel and household linen

- Source
- Vegetable or Cellulosic
- Animal
- Jute
- Linen
- Wool
- Fibre's Name
- Natural
- or Proteinic
- Cotton
- Silk
- Man-Made

## FABRIC STRUCTURE

Fabric structures are architecturally innovative forms of constructed fibers that provide end users a variety of aesthetic free-form building designs. Custom-made fabric structures are engineeredand fabricated to meet worldwide structural, flame retardant, weather-resistant, and natural force requirements. A fabric structure's material selection, proper design, engineering, fabrication, and installation are integral components to ensuring a sound structure.

## FABRIC STRUCTURE BASICS

### Membrane Materials

Most fabric structures are composed of actual fabric rather than meshes or films. Typically, the fabric is coated and laminated with synthetic materials for increased strength, durability, andenvironmental resistance. Among the most widely used materials are polyesters laminated or coated with polyvinyl chloride (PVC), and woven fiberglass coated with polytetrafluoroethylene(PTFE).

### Cotton Canvas

The traditional fabric for fabric structures in light cotton twil, light canvas, or heavy proofed canvas.

### Polyesters

Strength, durability, cost, and stretch make polyester material the most widely used in fabric structures. Polyesters that are laminated or coated with PVC films are usually the least expensive option for longer-term fabrications.

Laminates generally consist of vinyl films over woven or knitted polyester meshes (called scrims or substrates), while vinyl-coated polyesters usually have a high-count, high-tensile base fabric coated with a bondable substance that provides extra strength. Our WSSL'S fabric supplier, Ferrari, places the polyester fabric under tension both before and during the coating process. This results in a weave that has increased dimensional stability and is known commercially as Precontraint.

### Vinyl-laminated Polyesters

A laminated fabric usually is composed of a reinforcing polyester scrim pressed between two layers of unsupported PVC film. For most fabric structure uses, however, it refers to two or more layers of fabric or film joined by heat, pressure, and an adhesive to form a single ply.

With an open-weave or mesh polyester scrim, the exterior vinyl films bond to themselves through the openings in the fabric. Heavier fabric scrims, however, are too tightly woven to allow the same bonding. In this case, an adhesive is used to bond the exterior films to the base fabric.

A good chemical bond is critical to both prevention of delamination and development of seam strengths. The seam is created when vinyl-coated fabrics are welded together.

The adhesive enables the seam to meet shear forces and load requirements for a structure at all temperatures. The adhesive prevents wicking of moisture into the scrim's fibers, which also prevents fungal growth or freezing that could affect the exterior coating's adhesion to the scrim. Adhesives are water-based to comply with EPA regulations.

Open-weave scrims generally make the fabric more economical, although this can also depend on the number and type of features that you require in the vinyl. Almost any colour, [[Vinyl coated polyester | UV resistance Vinyl coated polyester]], and colorfastness may be incorporated into the vinyl. However, the more features added, the higher the cost of the fabric.

### Vinyl-coated Polyester

Vinyl-coated polyester is the most frequently used material for flexible fabric structures. It is made up of a polyester scrim, a bonding or adhesive agent, and exterior PVC coatings. The scrim supports the coating (which is initially applied in liquid form) and provides the tensile strength, elongation, tear strength, and dimensional stability of the resulting fabric. Vinyl-coated polyester is manufactured in large panels by heat-sealing an over-lap seam with either a radio-frequency welder or a hot-air sealer. A proper seam will be able to carry the load requirements for the structure. The seam area should be stronger than the original coated fabric when testing for tensile strength.

The base fabric's tensile strength is determined by the size (denier) and strength (tenacity) of the yarns and the number of yarns per linear inch or meter. The larger the yarn and the more yarns per inch, the greater the finished product's tensile strength. The adhesive agent acts as a chemical bond between the polyester fibers and the exterior coating and also prevents wicking, or fibers absorbing water, which could result in freeze-thaw damage in the fabric.

The PVC coating liquid (vinyl Organisol or Plastisol) contains chemicals to achieve the desired properties of colour, water and mildew resistance, and flame retardancy. Fabric can also be manufactured that contains high levels of light transmission or can be made completely opaque. After the coating has been applied to the scrim, the fabric is put through a heating chamber that dries the liquid coating. PVC coatings are available in a range of colours, although non-standard colours can be pricey. Colours may be subject to minimum order runs that allow the coating machine to clear out traces of any previous colour.

**Fiberglass**

Woven fiberglass coated with PTFE (Teflon or silicone) is also a widely used base material. Glass fibers are drawn into continuous filaments, which are then bundled into yarns. The yarns are woven to form a substrate. The fiberglass carries a high ultimate tensile strength, behaves elastically, and does not suffer from significant stress relaxation or creep. The PTFE coating is chemically inert, can withstand temperatures from 100°F upwards to 450°F+. It is also immune to radiation and can be cleaned with water.

Because of its energy efficiency, high melting temperature and lack of creep, fiberglass-based fabrics have been the material of choice for stadium domes and other permanent structures, particularly in the United States. However, when properly constructed, polyester structures may be equally durable.

**Olefin/ polyolefin**

A number of polymers consisting mainly of polyethylene, polypropylene or combinations of the two are available for fabric structures.

**PVDF Woven**

PVDF woven fabric are available for fabric structures.

**ePTFE Woven**

ePTFE woven fabric are available for fabric structures.

**Blackout Fabric**

Blackout material, also known as blockout material, is an opaque fabric. Blackout fabric consists of a laminate that sandwiches an opaque layer between two white exterior layers. Heating and lighting of a structure may be controlled because the fabric does not allow light to permeate the top or walls. The opaque quality also prevents stains, dirt, repairs, or slightly mismatched panels on the structure's exterior from being noticed from the inside.

**Topcoatings**

Most fabrics used for fabric structures have some form of topcoating applied to the exterior or coating to make cleaning easier. Topcoating provides a hard surface on the outside of the material, forming a barrier that aids in preventing dirt from sticking to the material, while allowing the fabric to be cleaned with water.

As the material ages, the topcoating will eventually erode, exposing the fabric to dirt and making it more difficult to clean. The thicker the topcoating, the longer it will last. However, coatings that are too thick will embrittle and crack when folded.

There are several commonly used topcoatings:

- PVF film lamination is made up of polyvinyl fluoride (commercially known as Tedlar). It consists of a film layer laminated to the PVC fabric during the manufacturing process. The result is a thicker fabric that can resist weather and chemical agents better than competing fabrics. Its self-cleaning ability repels such things as acid rain, graffiti, and bird droppings. Because of these characteristics it can often be found in use in industrialized areas, desert regions, and coastal zones. The thicker coating has a slow erosion rate, resulting in a life expectancy of approximately 25 years depending upon environmental conditions. This topcoating is flexible, which creates a strong and consistent bond to the PVC. PVF coated fabrics may be manufactured in a variety of colours, but are also subject to minimum manufacture runs. PVF coating also makes the fabric non-weldable. Seams are not overlapped, but instead are butted with an extra welded seam, then applied to the underside of the fabric that does not have a topcoating.
- Acrylic topcoating is the most economical and widely available topcoating used. Its spray-on application provides a glossy finish and a resistance to UV degradation. Because the coating is thin, the material is both easy to fabricate and is reparable by high frequency or hot air welding. Depending on environmental conditions, an acrylic topcoating will give the fabric a life span of 10 years or more. Acrylic topcoats are ideal for fabric structures, and can be found on products

used as tradeshow pavilions, temporary concert halls, and portable warehouses.

- PVDF topcoating is composed of a mix of fluorine, carbon, and hydrogen. The combination of the fluorine and carbon provides superior resistance to UV degradation and chemical damage than that of the acrylic topcoat. PVDF coated fabrics also maintain colour longer than do those coated with acrylic. PVDF fabrics resist algae and fungi, and also have self-cleaning properties that make them easy to maintain. These coated fabrics are flexible, resist cracking, are easily handled, and generally have a life span of 15 to 20 years depending upon environmental conditions. PVDF is chemically grafted to the PVC and polymers used, which reduces the colour choices. Because white is the only standard colour, other colour choices are limited and must be specially manufactured. Because of the chemical properties of the coating, it must be abraded off to expose the PVC before welding. This, too, increases fabrication costs. On site repairs are also difficult, as the membrane must be manually abraded before repair.
- PVDF/PVC topcoating is essentially a dilution of the PVDF topcoat that makes the finished fabric less expensive to manufacture. The finished fabric may be welded without having to abrade the material, which reduces costs. However, because the PVDF is diluted, longevity and the ability to withstand environmental factors are reduced. This results in a general life expectancy of 10 to 15 years. These materials are also available in colours, but are subject to limited manufacturing runs.
- Tio2 (Titanium Dioxide) Top coating "Titan W": TITAN W is a surface finish whose formulation was prepared by Naizil S.p.A. in collaboration with the Chemical Engineering Department of Padua University.

TITAN W exploits the most innovative nanotechnology methods for the use of particular particles in combination with fluoropolymers (PVDF) mixed with acrylic resin. The effectiveness of TITAN W compared to coatings based on acrylic lacquering and PVDF has been shown by numerous tests on deterioration by atmospheric agents, both accelerated and outdoors (weathering test).

The result is a coating that gives the fabric the following advantages: - Less yellowing and colour variation - Longer lasting physical properties - Better preservation of flexibility - Better self-cleaning properties. With these additional advantages, the fabrics coated with TITAN W lacquering will offer better durability compared to the existing coatings used. In addition to the above-mentioned characteristics, fabrics with TITAN W lacquer are high frequency and hot air weldable.

### Fabric Properties

When discussing fabric properties for use on a structure, there are several terms that are commonly used:

- Tensile strength is a basic indicator of relative strength. It is fundamental for architectural fabrics that function primarily in tension.
- Tear Strength is important in that if a fabric ruptures in place, it generally will do so by tearing. This can occur when a local stress concentration or local damage results in the failure of one yarn, which thereby increases the stress on remaining yarns.
- Adhesion strength is a measure of the strength of the bond between the base material and coating or film laminate that protects it. It is useful for evaluating the strength of welded joints for connecting strips of fabric into fabricated assembly.
- Flame retardancy does not have the same meaning as flameproofing. Fabric that contains a flame-retardant coating can withstand even a very hot point source. However, it can still burn if a large ignition source is present.

### Structural Properties

When deciding on a fabric it is imperative to keep certain fabric properties in mind. These include stress versus strain (unit load versus unit elongation), expected service life, the mechanisms of joining the material together (welding, gluing, etc.), and the fabric's behaviour in or around fire.

Stress versus strain data should be obtained in both uniaxial and biaxial forms. This information characterizes the fabric in terms of stiffness, elasticity, and plasticity. This is essential information when determining the material's response under load in a load-carrying application. Shear strength, shear strain, and Poisson's ratios, though difficult to obtain, are fundamental when analyzing a fabric as a structural material.

## COST SAVINGS TO FABRIC BUILDINGS

The benefits of fabric buildings are many - No lighting is required as fabric is translucent, which makes it an energy efficient solution. Mobility: You can move them, either on wheels or relocate them completely. Savings: They cost about half of what a traditional structure costs.

Fabric properties When discussing fabric properties for use on a structure, there are several terms that are commonly used: Tensile strength is a basic indicator of relative strength. It is fundamental for architectural fabrics that function primarily in tension. Tear Strength is important in that if a fabric ruptures in place, it generally will do so by tearing. This can occur when a local stress concentration or local damage results in the failure of one yarn, which thereby increases the stress on remaining yarns.

Adhesion strength is a measure of the strength of the bond between the base material and coating or film laminate that protects it. It is useful for evaluating the strength of welded joints for connecting strips of fabric into fabricated assembly. Flame retardancy does not have the same meaning as flameproofing. Fabric that contains a flame-retardant coating can withstand even a very hot point source. However, it can still burn if a large ignition source is present.

Of course, other properties must be factored in when determining a material's suitability for a structure. To fully understand a fabric's value and usefulness, consider the following:

- Shading coefficients
- General solar, optical, and thermal performance data
- Acoustical data
- Dimensional stability
- Colorfastness
- Cleanability
- Seam strength and stability
- General handling ability things like:including abrasion resistance, foldability, etc.

## FEATURES OF DIFFERENT KIND OF FIBRES

The properties of a fabric depend mainly on the properties of the fibre or fibres; from which they are made. Hence familiarity with the characteristic properties of a fibre will aid the consumer in knowing what to' expect of various fabrics an what care is needed in maintaining them. Table summarizes the properties of some of the fibres.

### Merits of Cotton

- Fabrics are easy to launder. They can also be drycleaned.
- Cotton fabrics can be readily dyed or printed with almost all the classes of dyes (barring acid and disperse). Pigment printing can be done with the help of a binder. Acid and basic dyes are not utilised on cotton, since cotton has no affinity for these dyes.
- Cotton can be processed into a wide selection of fabrics. Among the sheer fabrics are cambric, batist chiffon, lawn, organdy and voile. Medium weight cottons comprise broadcloth, drill, flannel, poplin, terry cloth and long cloth. The heavy weight cottons include brocade, denim, corduroy pique and between.
- Cotton can be mercerized to improve lustre, absorbency, dyeability and strength.
- Cotton fabrics can be resin finished to give them wrinkle resistance.
- Textured effects are achieved either by yarn structure (high twist

yarns, nub yarns, boucle yarns, crimped yarns), fabric construction (crepes, seersuckers) or application of special finishes (embossing, napping) conductor of heat and hence is very comfortable to wear.
- The wear-life of cottons is excellent as the fibre is strong.
- Various treatments like water repellent finishes, fire retardent and mildew resistant finishes can be provided to cottons.

**Limitations**

- Untreated cotton wrinkles easily.
- Cotton is susceptible to mildew if left damp.
- Cotton is flammable.
- Linen is very durable. It is two to three times stronger than cotton.
- Fabrics take a long time to dry.

**Linen Merits**

- It can be easily laundered as it is stronger when wet and has a good resistance to alkalies and bleaches. Linen, like cotton, can be easily dyed or printed. 5. Linen withstands high ironing temperatures.
- It is comfortable to wear because of its high absorbency, smoothness and good conduction of heat./"
- Linen is flammable.
- Linen wrinkles easily.
- It lacks flex abrasion and hence shows wear on crease lines and seams.
- Linen has poor press-crease retention.

**Demerits**

- Linen is susceptible to mildew if left damp.
- Fabrics take long time to dry.

**Advantages**

- It can be dyed with all the dyes that are used to cotton. It even has affinity for basic dyes.
- Jute can be treated with caustic soda to give it a wool-like appearance.
- It is a cheap fibre. It can be mixed with other fibres for economic and aesthetic reasons.
- It can be given finishes to develop its durability, its resistance to rot, water, fire and microbes.

**Demerits of Jute**

- It is harsh to the touch.
- It rots and is susceptible to microbial decomposition.
- It is a weak fibre.
- It is not easy to bleach.

- It sheds lint.
- It is difficult to wash or dryclean.

**Wool Merits**

- Fabrics oppose wrinkling as the fibre is very resilient and elastic.
- Because of their elasticity, woollen garments, specially knit wear, are comfortable as they yield to body movements.
- Fibres and fabrics are water repellent and fire resistant.
- Wool has excellent absorbency which makes it comfortable for the wearer.
- It provides warmth.
- Wool can be given anti shrink and moth resistant finishes.
- Wool dyes easily particularly with acid dyes in a wide range of colours.
- It is cool in lightweight fabrics.
- Wool may be laundered or drycleaned.

**Demerits**

- Wool tends to stretch and lose its shape when wet.
- Woollen garments tend to stretch during wear.
- Wool tends to build-up static electricity. Silk
- Wool is very sensitive to alkaline substances.
- Some people are allergic to wool. It led to skin rash and irritation.
- Wool felts in moist heat and friction.
- Wool is attacked by moths and carpet beetles.
- Wool needs special handling in laundering or drycleaning to prevent
- Chlorine bleaches affect wool adversely, felting, shrinkage or stretching out of shape.

**Merits**

- Fibres are stronger than wool. Silk is the strongest natural fibre.
- Silk fabrics have lustre and a soft feel.
- Silk can be constructed into a variety of fabrics; very sheer drapable J fabrics like georgette, chiffon, china silk, crepe de chine or heavy, stiff fabrics suchas, broadcloth, brocade, tafetta, faille, grosgrain, ottoman and velvet.
- Silk is warm and suitable for winter.
- Silk is absorbent and hence comfortable to wear.
- Fabrics are resilient and elastic and therefore resist wrinkles to some extent.
- White silk can be bleached with hydrogen peroxide or sodium perborate.
- Silk can be laundered or drycleaned depending on the dye and the finish.

- Silk can be easily dyed and printed to provide beautiful, brilliant shades.

**Demerits**

- Silk is sensitive to alkalies and to some acids.
- Sunlight affects silk adversely.
- Sunlight and perspiration weaken the fabric.
- Insects like carpet-beetle and silver-fish attack silk.
- Some silk fabrics leave water spots.
- Silk yellows with age.
- Silk is damaged by chlorine bleaches.
- Several dyes used on silk are affected by sunlight and perspiration, some are not fast to washing.
- Strong alkaline soaps and high temperatures tend to weaken and yellow silk.

**Merits**

- Viscose rayon can be dyed easily with all the dyes that take on to cellulosic fibres.
- Viscose rayon is suitable for a wide range of fabrics from lightweight luxury kinds that drape well to heavy, strong, durable types that are stiff and crisp.
- It can be provided a variety of finishes.
- It can be bleached with chlorine bleaches but peroxide bleach is preferable.
- Viscose rayon blends well with other fibres.
- It can be laundered or drycleaned, depending on the dye used and/or the finish given.
- It is very absorbent.

**Demerits**

- Fabrics have very poor dimensional stability due to progressive shrinkage of the fibres.
- Rayon is susceptible to mildew.
- Rayon is damaged more easily than cotton by mineral acids
- Fibres are weaker than cotton and weaker still when wet.
- Abrasion resistance is poor.
- Fabrics wrinkle easily because of the poor resiliency of the fibres.
- Fabrics require low ironing temperature.

**Merits of Palynosic Fibres**

- Abrasion resistance is better than that of viscose rayon.
- Fabrics are absorbent.

- Fabric possesses better strength than viscose rayon.
- Fabrics may be crisp and firm with good drapability.
- Dimensional stability of the fabric is good.
- Their lustre is subdued resembling that of silk.
- Polynosics can be given resin treatment for wrinkle resistance without any loss in strength.
- Fibres can be dyed and printed like cotton.
- Fabrics are easy to launder and can also be drycleaned.

**Demerits**

- Fabrics need low ironing temperatures.
- Fibre has poor resiliency.
- Fabrics are weakened by sunlight.
- Polynosics are susceptible to mildew.

**Merits of Acetate**

- Fabrics have a luxurious soft feel and silky appearance.
- Fabrics are unaffected by mildew or moth.
- Since the fibres are thermoplastic durable surface effects like embossing, schreinering, moireing, cireing can be given to the fabric.
- Acetate can be bleached with hydrogen peroxide or sodium perborate at low temperatures.
- Acetate has great draping qualities.
- Acetate can be dyed with disperse dyes or it can be dope dyed. Dope dyed fibres have excellent fastness to light, crocking, perspiration, washing and gas-fading.

**Demerits**

- Fabrics build up static electricity.
- Fibre dissolves in acetone and acetic acid.
- Acetate requires special dyes.
- Acetate fabrics have poor absorbency.
- They are weak when wet.
- Fibre weakens on exposure to sunlight.
- Acetate has poor abrasion resistance.

**Merits of Nylon**

- Nylon is unaffected by alkalies.
- Nylon has excellent abrasion resistance.
- Fabrics are dimensionally stable.
- Nylon is very strong even when wet.
- Nylon is resilient and elastic.
- Nylon washes easily and dries quickly.
- Fabrics need little or no ironing.

- The fabric may be either laundered or drycleaned and can be bleached or treated with fluorescent brighteners.
- Fabrics are unaffected by mildew and moths.
- Nylon resists water-borne stains.
- Fabrics can be heat-set to retain pleats.

**Demerits**

- Nylon melts in fire.
- Fabrics made from staple fibre tend to pill. This mars the surface appearance.
- Low ironing temperature is required.
- Fabrics are damaged by sunlight.
- Nylon fabrics build up static electricity.
- White nylon tends to pick up dyes and soils in laundering.
- Nylon fabrics absorb oil and oily soils. Oil stains are difficult to remove.

**Merits of Polyester**

- Polyester has high abrasion resistance.
- Fabric can be heat-set to retain pleats or creases.
- Fabrics have excellent wash and wear features. They show smooth.
- Fabrics are very strong.
- Polyester can be laundered or drycleaned.
- Polyester fabrics are wrinkle resistant.
- Fabrics have good dimensional stability.
- Polyester fibre can be texturized drying properties.
- Polyester blends well with other fibres to increase the wrinkle resistance,
- Polyester can be dyed with disperse dyes.
- Fabrics are resistant to moth and mildew.
- Polyester can be bleached of the other fibre and its wash-and-wear properties.

**Demerits**

- They have poor affinity for dyes.
- They have an affinity for oily soil and oil borne stains.
- Polyester can withstand moderate ironing temperatures.
- Polyester staple fibre fabrics tend to pill.
- They build up static charge.
- Polyester melts in fire.
- They have low absorbency.

**Merits of Acrylic**

- Fabrics are light in weight and have a pleasant feel.

- Fibres have good bulking properties abncs have good wrinkle resistance.
- Apparel can be easily laundered or drycleaned.
- Fibres show warmth characteristics similiar to that of wool.
- Fibres are resistant to acids, alkalies and bleaches.
- Fibres have good resistance to weathering and microbial attack.

**Demerits**

- Fabrics may pill relying on yarn construction.
- Fibres melt and decompose in fire.
- Fabrics have low absorbency.
- Acrylics and modacrylies are prone to retaining oil stains.
- Fabrics build up static electricity.

Varied Function of as seen above each fibre has its own merits and limitations. Textile scientists and technologists have worked out two ways, namely mixing of fibres and application of finishes, by which many disadvantages of the various textile fibres can be overcome. These are mixed fibre fabrics. Mixed fabrics are those made of yarns in which-two or more fibres are mixed while the yarn is spun. These yams may be used either in the warp or the weft or in both directions. In a union fabric each yarn is of a single class of fibres. Warp yarns may be of one kind of fibres e.g. cotton and the weft yarns of another e.g. wool. There may also be two or more kind of yams in each direction. In other words, in a *blend,* fibres are mixed beforeJ spinning while in *union fabrics,* fibres are mixed during weaving or plying of yarns. Whether the fabric is a *blend* or a *union fabric* the desirable properties of the constituent fibres will counterbalance their undesirable properties.

Take the case of *Terycot,* a polyester/cotton mix. Cotton has high absorbency but very poor wrinkle resistance. On the other hand, polyester has very low absorbency but has excellent wrinkle resistance. A mixture of 67/33 polyester/cotton shows good absorbency and good wrinkle resistance. A critical blend ratio is need to assure the best qualities of both the fibres. Better performance viz. better absorbency, durability, warmth, are not the only reasons for producing fabric blends and union fabrics. Improved aesthetics, drape, lustre, texture, colour and design effects are another reason why blends are manufactured. Economic reasons such as cost reduction, extension of fibre, such as, supplementing short supply, also leads manufacturers towards this. However, what we, as consumers need to understand here is that, in blend, the qualities of each fibre balance and supplement each other.

**NYLON, ACRYLIC**

## FINAL TOUCH

This is another way of introducing certain characteris-tics, one desires, into fabrics. Life has become very hectic today and most people cannot afford to Synthetic fibre fabrics gained popularity due to their easy care properties. However these fabrics are not as comfortable to wear as natural fibre fabrics, as they are not very absorbent and tend to build up static electricity. Natural fibre fabrics, in particular cotton, are being given certain chemical finishing treatments in order to help them compete with synthetic fibre fabrics.

**Laundry**

Clean, fresh, hygienic clothes, spotless bed-linen, immaculate towels, crisp and sparkling curtains are the outcome of successful laundering or drycleaning. Laundry is both an art and a science. It is a science in as much as it is based on I the application of scientific techniques. But it is also an art since the application of scientific techniques require the mastering of certain skills. Laundry includes two different types of processes: laundering and drycleaning.

Laundering is a process in which soils and stains are removed from textiles in an aqueous medium while drycleaning accomplishes these tasks in a non-aqueous medium. Fundamentally both these processes involve immersion of the textile in a solvent and agitation in it, in order to leach out the soil. Laundering covers a number of processes: washing, bleaching, drying any-ironing. It may also include pretreatment, stain removal and starching or softening. Laundry science helps carry out all these processes effectively with a minimum of wear and tear.

**Cellulosics**

***Cotton, Linen, Rayon and their Blends***

Chemical treatment used to prevent build-up of static electricity, which causes clinging of garment to the wearer and also attraction of soil. Chemical treatment which makes wool resistant to attack by moth and carpet beetles. Chemical treatment to prevent fabric from supporting flame. Application of resins to provides gloss,'smoothness, stiffness and/or embossed effects. Treatment with sodium hydroxide. Adds strength, lustre, absorbency. Improves dyeability.

Chemical treatment which helps retard or prevent development of mildew and mould in hot, humid climates. Mechanical or chemical finish to eliminate or minimize shrinkage. Mechanical finish meaking sure less than 1 per cent, shrinkage. Chemical treatment to control felting and shrinkage of wool. Treatment with gum, glue, starch, or resins to give smoothness, stiffness and strength.

Resin treatment which protects fabrics from stains and facilitates removal of oily soils. Chemicals, resins applied to make fabrics water repellent and resistant to water-borne soils. Treatment with resins to prevent excessive wrinkling or to impart drip-dry features or to produce durably pressed garments, permanent creases or pleats. Shape-retention is also improved. There are so many factors which govern *our* choice between laundering *or* drycleaning *of* a textile, the main one being the type *of* fibre *or* fibres *of* which the fabric is made. The properties *of* fibres that govern launderability are indicated in Table. Most fabrics can be either drycleaned *or* laundered without being adversely affected.

However certain fabrics, due to the fibre content, the finish, the fabric construction *or* the dye used may have to be only drycleaned, while others cannot be drycleaned because they include a component such as a bonding agent *or* a resin, *or* a dye that would be adversely affected by the drycleaning solvent. Individual properties that influence the manner in which a garment *or* household item is to be cleaned are:

***Cont***

A 100 per cent cotton fabric can withstand frequent laundering and rough handling when wet. Whereas a 100 per cent wool fabric has to be washed gently. A 100 per cent polyester *or* polyamide fabric cannot be ironed with the iron at the same temperature as a pure cotton fabric.

***Yarn Construction***

Fabrics *of* the same fibre content but *of* different yarn construction possess various qualities. Fabrics made from filament yarn, staple yam, carded *or* combed, simple yam *or* fancy yarn, loose twist *or* tight twist all perform differently.

***Construction of Fabric***

Various knitted fabrics may necessitate being handled differently. Lace fabrics may have to be treated with much greater care than woven fabrics. Woven fabrics with long floats also required careful handling.

***Final Touch***

Today many different finishes, physical and chemical are applied to fabrics. These have to be taken into view. Fabrics with chlorine retentive finishes suchas resin finishes, cannot be treated with chlorine bleaches as they would yellow.

***Properties of Dyes***

The fastness properties *of* dyes require much attention. Coloured clothes if they are not *of* guaranteed fast colour should be washed separately as these would bleed and could stain the other clothes. The may *of* drying coloured garments will differ. They are generally to be dried in the shade *or* if exposed to sunlight they need to be turned inside *out* and also removed as soon as dry.

## MINERAL WOOL FIBRE

Mineral wool has a unique range of properties combining high thermal resistance with long-term stability. It is made from molten glass, stone or slag that is spun into a fibre-like structure which creates a combination of properties that no other insulation material can match. It has the ability to save energy, minimize pollution, combat noise, reduce the risk of fire and protect life and property in the event of fire.

*Mineral wool: Production and properties* describes the technological process of mineral wool production and the physical characteristics of the melt and theoretical bases of multiregression and dimensionless theory. This is followed by the introduction of the fibre cooling model in the blow-away flow and the influence of temperature in the melt film (on the rotating centrifuge wheels) on the thickness of forming fibres.

The second part predominantly focuses on the use of computer-aided visualisation: tools for the diagnostics of fibre and primary layer formation. Special attention is given to the study of aerodynamic characteristics of the airflow which significantly influences the quality of the final product.

*Mineral wool: Production and properties* is suitable for engineers, researchers and for graduate and postgraduate students who want to broaden their knowledge of experimental methods in this field.

### ASBESTOS

Asbestos is a mineral fiber that has been used commonly in a variety of building construction materials for insulation and as a fire-retardant. Because of its fiber strength and heat resistant properties, asbestos has been used for a wide range of manufactured goods, mostly in building materials (roofing shingles, ceiling and floor tiles, paper products, and asbestos cement products), friction products (automobile clutch, brake, and transmission parts), heat-resistant fabrics, packaging, gaskets, and coatings.

When asbestos-containing materials are damaged or disturbed by repair, remodeling or demolition activities, microscopic fibers become airborne and can be inhaled into the lungs, where they can cause significant health problems.

*Most Common Sources of Asbestos Exposure*:

- Workplace exposure to people that work in industries that mine, make or use asbestos products and those living near these industries, including:
  - The construction industry (particularly building demolition and renovation activities),
  - The manufacture of asbestos products (such as textiles, friction products, insulation, and other building materials), and
  - During automotive brake and clutch repair work

- Deteriorating, damaged, or disturbed asbestos-containing products such as insulation, fireproofing, acoustical materials, and floor tiles.

## Types and Associated Fibres

Six minerals are defined by the United States Environmental Protection Agency as "asbestos" including that belonging to the serpentine class chrysotile and that belonging to the amphibole class amosite, crocidolite, tremolite, anthophyllite and actinolite.

There is an important distinction to be made between serpentine and amphibole asbestos due to differences in their chemical composition and their degree of potency as a health hazard when inhaled.

### *Serpentine*

*White*

Chrysotile, CAS No. 12001-29-5, is obtained from serpentinite rocks which are common throughout the world. Its idealized chemical formula is $Mg_3(Si_2O_5)(OH)_4$. Chrysotile fibers are curly as opposed to fibers from amosite, crocidolite, tremolite, actinolite, and anthophyllite which are needlelike. Chrysotile, along with other types of asbestos, has been banned in dozens of countries and is only allowed in the United States and Europe in very limited circumstances. Chrysotile has been used more than any other type and accounts for about 95% of the asbestos found in buildings in America. Applications where chrysotile might be used include the use of joint compound. It is more flexible than amphibole types of asbestos; it can be spun and woven into fabric. The most common use is within corrugated asbestos cement roof sheets typically used for outbuildings, warehouses and garages. It is also found as flat sheets used for ceilings and sometimes for walls and floors. Numerous other items have been made containing chrysotile including brake linings, cloth behind fuses (for fire protection), pipe insulation, floor tiles, and rope seals for boilers.

### *Amphibole*

*Brown*

Amosite, CAS No. 12172-73-5, is a trade name for the amphiboles belonging to the *Cummingtonite - Grunerite* solid solution series, commonly from Africa, named as an acronym from Asbestos Mines of South Africa. One formula given for amosite is $Fe_7Si_8O_{22}(OH)_2$. It is found most frequently as a fire retardant in thermal insulation products and ceiling tiles.

*Blue*

Crocidolite, CAS No. 12001-28-4 is an amphibole found primarily in southern Africa, but also in Australia. It is the fibrous form of the amphibole riebeckite. One formula given for crocidolite is $Na_2Fe^{2+}{}_3Fe^{3+}{}_2Si_8O_{22}(OH)_2$. Notes: chrysotile commonly occurs as soft friable fibers. Asbestiform amphibole may also occur as soft friable fibers but some varieties such as amosite are commonly straighter. All forms of asbestos are fibrillar in that they are composed of fibers with widths less than 1 micrometer that occur in bundles and have very long lengths. Asbestos with particularly fine fibers is also referred to as "amianthus". Amphiboles such as tremolite have a crystal structure containing strongly bonded ribbonlike silicate anion polymers that extend the length of the crystal. Serpentine (chrysotile) has a sheetlike silicate anion which is curved and which rolls up like a carpet to form the fibre.

*Other Materials*

Other regulated asbestos minerals, such as tremolite asbestos, CAS No. 77536-68-6, $Ca_2Mg_5Si_8O_{22}(OH)_2$; actinolite asbestos, CAS No. 77536-66-4, $Ca_2(Mg, Fe)_5(Si_8O_{22})(OH)_2$; and anthophyllite asbestos, CAS No. 77536-67-5, $(Mg, Fe)_7Si_8O_{22}(OH)_2$; are less commonly used industrially but can still be found in a variety of construction materials and insulation materials and have been reported in the past to occur in a few consumer products.

Other natural and not currently regulated asbestiform minerals, such as richterite, $Na(CaNa)(Mg, Fe^{++})_5(Si_8O_{22})(OH)_2$, and winchite, $(CaNa)Mg_4(Al, Fe^{3+})(Si_8O_{22})(OH)_2$, may be found as a contaminant in products such as the vermiculite containing zonolite insulation manufactured by W.R. Grace and Company. These minerals are thought to be no less harmful than tremolite, amosite, or crocidolite, but since they are not regulated, they are referred to as "asbestiform" rather than asbestos although may still be related to diseases and hazardous.

## BASALT FIBER

Basalt fiber or fibre is a material made from extremely fine fibers of basalt, which is composed of the minerals plagioclase, pyroxene, and olivine. It is similar to carbon fiber and fiberglass, having better physicomechanical properties than fiberglass, but being significantly cheaper than carbon fiber. It is used as a fireproof textile in the aerospace and automotive industries and can also be used as a composite to produce products such as camera tripods.The manufacture of basalt fiber requires the melting of the quarried basalt rock to about 1,400 °C (2,550 °F). The molten rock is then extruded through small nozzles to produce continuous filaments of basalt fiber. There are three main manufacturing techniques, which are centrifugal-blowing, centrifugal-multiroll and die-blowing. The fibers typically have a filament diameter of between 9 and 13 $\mu$m which is far enough above the respiratory limit of 5 $\mu$m to make basalt fiber a suitable replacement for asbestos. They also have a high elastic modulus, resulting in excellent specific tenacity - three times that of steel.

## GLASS WOOL

Glass wool is an insulating material, made from fibre glass, arranged into a texture similar to wool. Glass wool is produced in rolls or in slabs, with different thermal and mechanical properties.After the fusion of a mixture of natural sand and recycled glass at 1,450 °C, the glass that is produced is converted into fibres. The cohesion and mechanical strength of the product is obtained by the presence of a binder that "cements" the fibres together. Ideally, a drop of bonder is placed at each fibre intersection. This fiber mat is then heated to around 200 °C to polymerize the resin and is calendered to give it strength and stability.

The final stage involves cutting the wool and packing it in rolls or panels under very high pressure before palletizing the finished product in order to facilitate transport and storage. Glass wool having better advantages compair to other insulation materials. But it is hazardous due to its duct & very small glass particles which can travel into the human body during inhalation. Glass wool is a thermal insulation that consists of intertwined and flexible glass fibres, which causes it to "package" air, resulting in a low density that can be varied through compression and binder content. It can be a loose fill material, blown into attics, or, together with an active binder sprayed on the underside of structures, sheets and panels that can be used to insulate flat surfaces such as cavity wall insulation, ceiling tiles, curtain walls as well as ducting. It is also used to insulate piping and for soundproofing.

## THE DEVELOPMENT OF MODERN FABRIC STRUCTURES

The simple dwellings tended to be fabricated in traditional ways by the intended occupants. As they degraded, their components were replaced, and so the overall design evolved. More complex fabric structures were predominately the craftsman's trade and not the domain of the architect. This was to change during the nineteenth and twentieth centuries as architects became inspired by technological breakthroughs instructural engineering, made more appealing by architectural theorising on the emerging functional aesthetic.

In 1823 at around the same time that strong ropes first became commercially available, Claude Navier published a seminal study on suspension structures. The teething problems of the resulting generation of long-span structures were substantially laid to rest in 1838 when J.M. Rendel identified torsional stability as the major cause of bridge collapse. Large suspension structures rapidly became a common feature of engineering design, and it was not long before their spanning potential was used for large buildings.

At the 1896 Nizny Novgorod exhibition in Russia, the Structural Engineering Pavilion designed by V.G. Shookhov consisted of a large radial steel cable net pre-tensioned in both directions and clad with steel panels. It was believed that a similar structure composed of thicker 3mm steel sheets could span over 300m. This provoked an interest in the spanning potential of tension stabilised surfaces which has continued to the present day.

A parallel development began in 1918 when F.W. Lanchester patented a design for an 'air tent' in which it was proposed that a patterned balloon fabric could be inflated at a low pressure to form a habitable enclosure. In 1938 Lanchester developed the concept further with a design for an air supported dome spanning over 650m. Such air supported and air inflated structures had many potential applications but in 1946 a mass market was identified, creating 'radomes', minimal structure shelters which provided climatic protection for radar dishes.

Following successful trials between 1946 and 1950, the radome concept was applied to the DEW line early warning system, and in 1953 Walter Bird established the company Birdair primarily for the manufacture of radomes. Since 1946 tens of thousands of radomes have been manufactured, some of the largest being more than 60m in diameter. This generated a great deal of interest in the whole subject of tensile surface structures and stimulated intensive research into fabric composition. Fabric structures had become big business.

In the same year that Birdair began trading, the L.S. Dorton Arena in Raleigh was completed. Designed by Matthew Norwicki and Fred Severud, it was the first modern, doubly curved, pre-stressed saddle structure. Whilst itself not actually made of fabric, its completion is popularly believed to signify the birth of the modern 'tent'. The Raleigh Arena influenced many architects, and directly inspired both Saarinen's Yale Hockey Rink (1956) and Tange's two stadia for the Tokyo Olympics (1961).

A young German architect with a personal interest in lightweight and tension structures visited America during this period, where he met both Saarinen and Severud. The architect, Frei Otto, began exhaustive investigations into the structural principles behind this new generation of buildings. Even before the publication of his thesis 'Das Hangende Dach' in 1954, the practical implications of his work were brought to the attention of the tent manufacturer Stromeyer and Co.

Peter Stromeyer made both his experience and resources available to Otto, and between them over the next twenty years, they were to undertake much of the intensive research and experimental construction which served to bring pre-stressed fabric structures into the vocabulary of the contemporary architect. In 1957 Frei Otto established the Development Centre for Lightweight Construction using money he had received for commissions, and in 1964 re-named the Institute for Lightweight Structures it became affiliated to the University of Stuttgart. The Institute was primarily concerned with developing methods for deriving the ideal forms for tensile surfaces. Initially these investigations were based on detailed studies of soap films and wire mesh models, however later Otto was to meet the mathematician Fritz Leonhardt, who set about making this form finding process more mathematically explicit. The variety of Otto's designs was wide. The first fabric structures he had built were a series of small music pavilions at the Federal Garden Exhibition at Kassel (1955), which were followed in 1957 by the similar twin saddle structures at the entrance to the Cologne Federal Garden Exhibition. He also designed convertible structures in the tradition of the Roman velarium, his first being the Theatre Terrace at Casino (1965).

Much of Otto's early research was embodied in the Montreal German Pavilion (1967), and a new permanence was heralded by the plexi glass clad cable net of the Berlin Olympic Stadium (1969/70).

In the late 70's the world recession depressed the market for air houses, and this was compounded by a series of severe storms which resulted in a number of highly publicised deflations. Fabric structures in general, and airhouses in particular, were considered contrary to the new 'long life/ low energy' construction ethic that was being advocated. This hindered the development of fabric structures, especially in Europe, and in 1974 Stromeyer's company went into receivership.

During the nineteen eighties and nineties, the fabric structures industry became more tentative, concerned with a confidence building process involving the consolidation of structural design techniques and the development of more reliable fabrics.

This however provided a sound base from which the size and complexity of fabric structures increased, to give us the extraordinary developments we now associate with modern fabric structures such as the fabric roof of the Hajj Terminal at Jeddah Airport (1981) covering nearly half a million square meters, the 320m diameter Millennium Dome (2000) and Denver International Airport.

## MATERIAL SELECTION OF FABRIC STRUCTURES

A fabric structure's material selection, proper design, engineering, fabrication and installation all work together to ensure a sound structure. The material's role in the structure's performance makes the selection process especially important. This is particularly true with tensile and air-supported structures because their membranes, as well as their frames, carry the loads.

## MEMBRANE MATERIALS

Most fabric structures use fabrics rather than meshes or films. The fabrics typically are coated and laminated with synthetic materials for greater strength and/or environmental resistance. Among the most widely used materials are polyester laminated or coated with polyvinyl chloride (PVC), woven fiberglass coated with polytetrafluoroethylene (PTFE) or silicone. Meshes, films and other materials also have appropriate applications.

## Polyesters

Polyester is the most frequently used base material because of its strength, durability, cost and stretch. Polyesters laminated or coated with PVC films generally are the least expensive for longer-term fabrications.

Laminates usually consist of vinyl films over woven or knitted polyester meshes (called scrims or substrates). Coated fabrics typically use a high-count, high-tensile base fabric coated with a bondable substance for extra strength. One fabric manufacturing method places polyester fabric under tension before and during the coating processes. The result is that yarns in both directions of the weave have identical characteristics, giving the fabric increased dimensional stability.

Lighter fabrics (200 to 270g/m$^2$) commonly are used as acoustic and insulated liners suspended beneath a structure's envelope. For long-term exterior use, heavier materials are needed: 20- to 26-oz. (680 – 880gm) fabrics with topcoatings of polyvinyl fluoride (PVF, of which Tedlar is an example) or polyvinylidene fluoride (PVDF, of which Vidar, Fluorex® and Kynar® are examples). These topcoatings provide a protective finish to withstand environmental degradation.

## Vinyl-coated Polyester

Vinyl-coated polyester is the most common fabric for producing flexible structures, such as custom-designed awnings, canopies, walkways, tent halls, smaller air-supported structures and light member-framed structures.

Vinyl-coated polyester is composed of a polyester scrim, a bonding or adhesive agent, and exterior PVC coatings. The polyester scrim supports the coating (applied initially in liquid form) and provides the tensile strength, elongation, tear strength and dimensional stability of the finished fabric. The scrim is made of high-tenacity, continuous-filament yarns, which have high dimensional stability, and can be bent thousands of times without losing any tensile properties.

The base fabric's tensile strength is determined by the size (denier) and strength (tenacity) of the yarns and the number of yarns per linear inch or meter. The bigger the yarn and the more yarns per inch, the greater the finished product's tensile strength. For architectural applications, base fabrics typically weigh between 2.5 and 10 oz/yd$^2$, with a tensile strength between 300 (2.662 N/5cm) and 650 lbs/in (5.60 N/5cm), although fabrics intended only for tent use may have lower measurements.

The adhesive agent provides a chemical bond between the polyester fibers and the exterior coatings and prevents wicking of moisture into the fibers. Wicking is the capillary like action of fiber to absorb water, which could result in freeze-thaw damage. The PVC coating liquid (vinyl Organisol or Plastisol) contains chemicals to achieve desired properties regarding colour, water resistance, mildew resistance and flame retardancy. The fabrics also can be made with high levels of light transmission or complete opaqueness. After the coating is applied to the scrim, the fabric goes through a heating chamber to dry the liquid coating.

**Vinyl-laminated Polyesters**

Vinyl-laminated polyesters are used for awnings, tents and low-tension frame structures. Technically, a laminated fabric consists of a reinforcing polyester scrim that is calendared between two layers of unsupported PVC film. In general use, it refers to two or more layers of fabric or film joined by heat, pressure and an adhesive to form a single ply.

With an open-weave or mesh polyester scrim, the exterior vinyl films bond to themselves through the openings in the fabric. Heavier base fabrics, though, are too tightly constructed to permit this lamination process, so an adhesive must bond the exterior films to the base fabric.

A good chemical bond is important to prevent delamination and is critical in developing the proper seam strengths. The adhesive enables the seam, created by welding vinyl-coated fabric to another piece of the same material, to meet a structure's shear forces and load requirements at all temperatures. By preventing wicking of moisture into the scrim's fibers, the adhesive prevents fungal growth or freezing that can affect the exterior coating's adhesion to the scrim. In response to EPA regulations, the adhesives are water-based.

Using an open-weave scrim such as mesh might make these fabrics more economical, depending on the number and type of features required in the vinyl. What weight is necessary to withstand abrasion and wear? Is flame resistance needed? Is a particular colour required? What width? Virtually any colour, plus UV resistance, abrasion resistance, and colorfastness can be formulated into the vinyl, but the more of these features incorporated, the higher the cost.

**Fiberglass**

Another widely used base material is woven fiberglass coated with PTFE (also known as Teflon®) or silicone. The glass fibers are drawn into continuous filaments, which are bundled into yarns. The yarns are woven to form a substrate. The fiberglass has a high ultimate tensile strength, behaves elastically and does not undergo significant stress relaxation or creep. The PTFE coating is chemically inert, withstands temperatures from minus 100F to 450F (minus 73C to 232C), is immune to UV radiation and can be cleaned with water.

PTFE-coated fiberglass is available with as much as 25% translucency, providing diffused interior light. Its ability to provide natural daytime lighting and its highly reflective surface for efficient nighttime interior lighting can reduce energy consumption. For these and other reasons, fiberglass-based fabrics have been the material of choice for stadium domes (both air- and cable-supported) and many other permanent structures, particularly in the United States. Another reason some industry experts cite for this is a perception among code officials that its high melting temperature and lack of creep, or long-term elongation, make it superior to polyester.

Other industry insiders note that polyester, like fiberglass, melts rather than burns at high temperatures, and that properly constructed, polyester structures may be equally durable.

Because of the differences in how polyester and fiberglass perform in fire-resistance tests, PTFE-coated fiberglass is the only membrane material that currently meets the U. S. model building codes definitions of a noncombustible material. (The three U.S. model codes are currently being reviewed and soon will be consolidated into one code.) This is a more accurate reason for the PTFE-coated fiberglass preference, but it raises questions about whether standards applied to other building materials should be applied to membranes.

### Woven PTFE

This material is constructed of PTFE fibers woven into a fabric. As of now, only one such material is available. Woven PTFE combines the environmentally-resistant advantages of the material with its ability to withstand repeated flexing and folding, an advantage it has over coated-fiberglass fabrics. Such flexibility makes it an especially good option for convertible structures; however, it is a rather expensive material and is not as strong as either polyester or glass.

### ETFE Foil

Perhaps the newest development in the fabric structures arena is the introduction of ETFE (ethylene tetrafluoroethethylene), a transparent membrane with fabric like qualities and the advantages of PTFE, such as a self-cleaning capability.

Resistant to atmospheric pollution and UV light, ETFE has a very long expected lifespan of more than 20 years. Effective thermal performance (average U value is 2.6W/m$^2$K for a two-layer system) and high light transmission (95% visible light, 85% UV light) enable a range of applications where traditional materials, such as glass, would not be practical. It is more than 20 times lighter than glass (0.35kg/m$^2$ for ETFE vs. 15kg/m$^2$ for glass) and is ecologically friendly and energy efficient as its constituent materials are fluorspar, hydrogen sulphate and trichloromethane, all non petrochemical derivatives. It is 100% recyclable.

### Meshes, Netting and Film

These are the least-used materials for fabric structures. Mesh is a broad term for any porous fabric with open spaces between its yarns. It can be made from almost any fiber by a variety of methods, including knitting, weaving and extrusion. In some cases it acts as a substrate to beef up other fabrics or is coated to produce specific characteristics. For architectural use, meshes typically are available as polyester weaves lightly coated with vinyl or as knitted fabrics using high-density polyethylene (HDPE), polypropylene or acrylic yarns. Polyester mesh dyes well, is strong, has a low water absorption rate and can be economical. Nylon often is used in industrial applications because of its strength and resistance to chemicals, although it does have a high water absorption rate and may cost more than polyester. Often used in agriculture, recreation and containment, polypropylene and HDPE are inert, so they can't be stained or dyed, and are less expensive than polyester or nylon. Polypropylene, however, does have a comparatively low melting rate, a factor in some industrial applications.

Meshes can provide shade as well as shelter from wind. Since they're porous, meshes are little good against rain. Still, they are inexpensive and have been used for some low-cost membrane structure applications. For obvious reasons, meshes are not used in traditional air-supported structure design.

Netting is considered a type of mesh, usually tight with small holes. Netting finds use in stadium interiors behind goals, golf ranges and courses, playground equipment and structures, horticulture, zoos, construction sites and other areas where protection or containment is needed.

Netting consists of a nylon, polyester or polypropylene with extruded or spun yarns that is knotted or raschel knitted to form the material. Each material has its advantages and appropriate applications. Polyester holds dye better than nylon but is more expensive; nylon is easier to coat, but has a higher water absorption rate and doesn't hold dye as well. Polypropylene floats on water, is durable and chemically resistant, but can't be dyed. Raschel knitting is a newer, faster manufacturing method than knotting. One drawback is that the knitted material can unravel, which can be thwarted by heat-setting the netting to shrink and stabilize the fibers.

Films are transparent polymers extruded in sheet form without a supporting substrate. They are not laminated or coated. Examples include clear vinyl, polyester or polyethylene. These films are cheaper than textiles, but they are neither as strong nor as durable.

Films are much weaker in tension, though more elastic, than scrim-based fabrics. Films sometimes have application in air-inflated structures. Air-inflated structures are composed of fabric tubes in which the air is pressurized, but the structure's interior itself is not.

Some air-inflated roofs or building envelopes have been made using two or three layers of films to form air pillows.

The film layers are thermally welded and sealed, and the resulting pillows are inflated by small fans. The inflation increases the internal pressure to prestress the surface, creating load resistance. Such film pillows are framed by an aluminum extrusion perimeter, which must accommodate some structural movement. Films range in thickness from 30 to 200 microns and can be produced with levels of translucency varying from 25% to 95% light transmission. Films are low weight, have a life expectancy of 20 to 25 years and highly resist dirt. The inflated pillows exhibit good thermal insulation values. More research needs to be done to develop a range of standard reliable, economical details, for instance, to improve the water seals and reduce wicking.

### Blackout Fabric

Blackout material, sometimes called blockout material, is an opaque or nontranslucent fabric. Primarily used for tents, the fabric is a laminate that sandwiches a dark opaque layer between two white exterior layers. Because no light transmits through the tent top or walls made with blackout material, lighting and heat can be controlled. In addition, any stains, dirt, repairs or slightly mismatched panels on the tent's exterior will not be noticeable from the inside.

Blackout fabric also has its disadvantages. Heating may be necessary, as the tent's interior may be colder than using nonopaque fabric. Heating and lighting, of course, will increase the operating costs. The multiple layers make the fabric heavy and harder to handle, and increase the possibility of delamination over time. The fabric is more expensive that other tent materials, pinholes or snags in it are apparent to occupants, and often, the interior colour does not exactly match the exterior colour.

## OTHER MATERIALS

Shadecloth, often a knitted fabric such as monofilament polyethylene, originally found use as agricultural crop protection. It has been adapted for tension structures whose purpose is solar shading. Shadecloth can be manufactured in a variety of colours, offers stretch and resiliency and remains flexible without tensile-strength loss under a range of conditions. Light transmission can vary from 20% to 90% shade factor; its UV filter construction can range from 30% to 70%.

Nylon, which may be laminated with vinyl films, is stronger and more durable than polyester, but has a higher cost and more stretch. It may be a good choice for some small jobs, but it stretches too much for use in large buildings.

Spandex is difficult to use in long-term exterior applications because of the wind's effect on its seams and construction, but its stretch and splash of colours can contribute to interior spaces or temporary exterior use, for instance, at festivals and special events.

Solution-dyed acrylic and modacrylic have gained use, particularly for small shading structures. Their wide range of colours and modacrylic's flame retardancy make the materials attractive to designers.

Kevlar® is an excellent lightweight fabric for construction, but it is very expensive, it is rarely used for large structures, such as dome roofs.

## FABRIC PROPERTIES

Product test data is almost the only way to establish a measure of relative quality. Many variables enter into the process of making fabrics, which may make one manufacturer's product significantly different from its competition's in one or more aspects. Test results provide the best indicator of such differences. Many suppliers have invested considerable money and time to test and characterize their products, and routinely provide information about the properties of their fabrics, including:

- Strip tensile strength
- Grab tensile strength
- Trapezoidal tear strength
- Tongue tear strength
- Adhesion strength
- Flame resistance
- Finished weight
- Base fabric weight
- Available topcoatings
- Resistance to cold cracking
- Dead load
- Structural properties
- Life expectancy

Tensile strength data is a basic indicator of relative strength. It's fundamental for architectural fabrics that function primarily in tension. Tear strength is important because if a fabric ruptures in place, it generally does so by tearing. This occurs when a local stress concentration or local damage results in the failure of one yarn, which increases the stress on remaining yarns.

Adhesion strength is a measure of the strength of the bond between the base material and coating or film laminate that protects it. The measure is useful for evaluating the strength of welded joints for connecting strips of fabric into fabricated assembly.

Flame retardancy is not the same as flame proofing. Fabric with a flame-retardant coating can withstand a point source even if it is very hot, but a flame-retardant material still will burn if a large ignition source is present. The larger the ignition source, the more total heat energy is available to the fabric fibers behind the protective coating, The more heat energy gets in, the faster and more successfully the fabric reaches a temperature at which it catches fire and burns from the inside out. Typical tent fires, for example, begin with small ignition sources, but ultimately the flammability of the tent's contents contributes to the fabric's response.

Flame-retardancy tests measure the self-extinguishing feature of fabric when subjected to a flame. The industry has developed AF-1 and AF-2 classifications for architectural fabrics. Both types must have a flame spread rating of 25 or less and provide at least a Class C roof covering. In addition, AF-1 fabrics must pass tests related to resistance to external fire exposure and interior flame spread. In certain temporary or nonbuilding structures, fabrics that meet NFPA 701 (flame resistance), or NFPA 701 in conjunction with a Class C classification, may suffice. Manufacturers should provide confirming information on which of the NFPA or ASTM tests their products pass.

## TOPCOATINGS

Most architectural fabrics have some form of topcoating applied to their exterior coating to improve cleanability. The topcoats are acrylic solutions, polyurethane-acrylic solutions, PVDF solution coats or a PVF film lamination. The topcoat provides a hard surface on the outside of the material and minimizes plasticizer migration. The barrier helps prevent dirt from sticking to the material and allows the fabric to be cleaned with water. As the material ages, the solution-coated top finishes will erode and the material will collect more dirt and be harder to clean. Thicker-solution topcoats last longer than thin coats, but coatings that are too thick will embrittle and crack when folded.

For permanent air structures and tensile structures, use of a 1-mil (25.4 microns) PVF film, particularly if long-term cleanability and appearance is an issue. The 1-mil PVF film is 10 times as thick as the solution topcoats and will eliminate plasticizer migration. The fabric's top finish should relate to the structure's long-term aesthetic requirements. Structures used for warehousing and industrial applications generally don't require high levels of cleanability. Air-supported structures for sports events, tennis courts or golf ranges require a moderate level of cleanability. Custom tensile structure for amusement parks and music pavilions generally require the highest level of cleanability.

## STRUCTURAL PROPERTIES

A fabric's most fundamental properties are related to stress versus strain (unit load versus unit elongation), expected service life, the mechanisms of joining the material together (welding, gluing, etc.) and the behaviour of the material in or around a fire. With this information, you are reasonably assured of being able to design a safe project.

For stress versus strain, data should be in the form of both uniaxial and biaxial information that characterizes the fabric in terms of its stiffness, elasticity and plasticity. The information is essential to effective modelling of the material's response under load in a load-carrying application. Shear strength, shear strain and Poisson's ratios are more difficult to obtain, but are fundamental for analyzing fabric as a structural material.

Fabric manufacturers should be able to provide evidence of the fabric's long-term performance in a representative environment based on testing aged samples.

Other properties come into play in evaluating a fabric's viability in a project. Finding information about these properties may be more difficult to obtain, but worth asking about to gain a full picture of the fabric's performance in a project. Some properties include:

- Shading coefficients
- General solar, optical, thermal performance data
- Acoustical data
- Dimensional stability
- Colorfastness
- Cleanability
- Seam strength and stability
- Construction method
- General handling ability, including abrasion resistance, foldability, etc.

## THE THERMOPLASTIC FIBRES

### NYLON

Nylon was the fibre first to be synthesized from materials none of which had previously been fibrous in nature.

*History.* The history of nylon dates back to 1928, when Dr. Stine. the Chemical Director of du Pont Company prevailed upon the Company to start research work relating to new fibres. The research continued for some time and experiments were made on cellulose derivations, particularly the esters arid new types of esters, and also on certain nitrogen containing derivatives of cellutose. These experiments were not successful Dr. Wallace H. Cenothen, at the same time made a study of poly-condensation whereby linear-polymers are produced, and this eventually led to the invention of nylon. In 1938, du Pont Company started a plant for nylon production. This fibre was known as fibre 66. Later, the name nylon was given to it.

### THE PROCESS OF MAKING NYLON

Although its often stated that nylon is made of coal, air and water, the actual synthesis is entirely different. Nylon 66 is produced from an acid and a diamine, which are produced from coaltar derivaties. A mixture of two coal-tar products, a dibasic acid, adipic acid, and hexamethylene diamine containing nitrogenis heated to give a condensed product known as nylon polymer. The, process is explained diagrammatically. Since each of the above compounds have six carbon atoms, the finished fibre made from this was known as Fibre 66.

The nylon polymer is melted and passed on chilled rollers when it comes out in the form of rolls, these are then cut into chips and stored

The chips are remelted and iltered through special filter packs.

After filtering, the molten polymer is passed through the tiny holes of a metal disc called a spinneret and fine filaments are thrown out into air where these get hardened. These are then passed through a conditioner which moistens them so that a number of these filaments stick together to form a long thread which is wound on a reel.

In order to increase the tensile strength and elasticity in the fibre, the filaments are drawn to about four times their original length by the application of force. During this operation the diameter of the fibre is reduced.

Twisting of the yarn is also carried on at the same time by the use of suitable equipment for the purpose.

## PROPERTIES OF NYLON

Nylon is a lustrous white transparent fibre and is both tough and pliable. Absorbency is low and this makes it a quick drying and easily washable fibre. Although it does not stain readily it tends to pick up colour, grease and soil if laundered with other garments.

It is not affected by cold temperatures but looses strength and yellows at sustained high temperatures. Pressing temperature of *24°F* is considered safe for nylon.

It has good wrinkle resistance and crease recovery. It has very good abrasion resistance and is considered a very durable fibre. It is degraded by exposure to light. It is not chemically reactive to soap, alkalies or alcohol but reacts readily to acid. Bleaches do not affect it and so:are ineffectual for whitening discoloured nylon.

They can be dyed and usually the colours are permanent. They are sometimes resin treated to stabilize the weave, silicone treated to improve softness or water-repellency but because of their low absorbency they do not accept finishes readily.

## TERRYLENE OR DACRON

The work of Carothers paved the way. For many other textile materials. Terrylene is one of these. Whinfield ard Dickson patented the method for making terrylene using teraphalic acid and ethylene glycol. This acid and alcohol are polymerized in vacuum at a high temperature and the polymer is extended in the form of a riboon. The polymers (molten) are passd through spinnerettes with circular perforations.

The Pont Company purchased patent rights of this fibre in the United States, and with some modifications in processing, put Dacron as a new fibre on the market in 1953. Dacron and terrylene are essentially the same fibre.

The most outstanding properties of this fibre are its excellent resistance to wrinkling and creasing, both when dry and wet. It has good abrasion resistance, toughness, resilience, elasticity and stretch resistance. Dacron has a property known as 'wicking'. Wicking is the ability of a fabric or fibre to pick up moisture and allow it to travel along the fibre topick up moisture and allow it to travel along the fibre, without actually absorbing it. Thus dacron is comfortable to wear in hot weather. Dacron melts and drops when exposed to fibre. It is self-extinguishing and does not flash burn. It has good resistance to weak acids and alkalies but is disintegrated by concentrated sulphuric acid. Dry cleaning solvents and bleaches do not effect this fibre. It is difficult to dye but can be satisfactorily colourel with acetate dyes.

## MATERIALS BASED ON NATURAL PROTEINS

*Ardil:* The raw material for the manufacture of ardil are the proteins, arachiil and conarchin which can be obtained from groundnuw.. It is a result of the research carried by Imperial Chemical Industry (I.C.I.).

*Fibrolane:* In principle the manufacture of fibrolane is similar to that of ardil, the main difference being that casein from milk is used instead of groundnut protein. People will be surprised if they are told that the suits they are wearing are of ilk instead of silk.

## MINERAL FIBRES

Three minerals are used in textiles fibre making; asbestos-a natural fibre; glass fibre and metallics a man-made fibre.

## ASBESTOS

Asbestos is a naturally occurring mineral fibre. It is fibrous rock and is famous for its non-burning qualities. Under the microscope the fibres appear straight, smooth and needle-shaped. It does' not burn or melt but short periods of high temperatures will decrease its strength. It is absorbent and has Yicking ability. It is acid and alkali resistance and this makes it specially desirable in the use of filters for chemicals and in industries where chemicals are used. Flameproof clothing for industrial and military purposes are also made from asbestos. It *is* used for all types of fire fighting equipment and insulation for steam pipes, brakes, and hundreds of other items where non-combustibility is essential.

## GLASS

'Fibreglas' is the trade mark for the glass fibre produced by the Owens-Corning Fiberglas Corporation of America and *is* die only Glass fibre about which information is available. Glass fibres are inorganic polymers based on silicon rather than carbon. These fibres have not become very popular as they are subject to severe abrasion and cut each other *if* they rub together. They are used for curtains and drapery fabrics because of their attractive appearance, non-combustibility and are resistant to light deterioration. They are not suitable for wearing apparel as the ends of these fibres irritate the skin, they are heavy and non-absorbent. Much of the glass fibre is used for insulation in buildings and for insulating electrical and other industrial equipment.

## METALLICS

Metallics are defined as "a manufactured fibre composed of metal, plastic-coated metal, metal-coated plastic, or a core completely covered by metal."

They are yarns and not fibres, but they are considered in fibre groupings because they replace fibres for textiles purposes. Gold and silver have long been used in India in textile fabrics. Aluminium and copper have a more recent origin. Metallics were used to add richness, glamour and glitte to fabrics. Modern metallic fibres, with aluminium as the basic metal, are much cheaper, softer and lighter in weight. But because they are no more rich and luxurious and look cheap this may lead to their rejection.

## SOME TRADITIONAL TEXTILES OF INDIA

The India of today succeeded to a rich heritage in Architecture, Music and other Fine Arts and Crafts, which for scores of centuries had been steadily growing despite the rise' and fall in the destinies of its people.

Aliens who came, whether as traders or as invaders, ended with the solitary exception of the English-by making India their home and by getting absorbed in its original inhabitants. The intermingling of races did start new currents and cross currents in the flow of our culture, but it made our culture richer and fuller. The culture, India invariably shared with its neighbours and others, thereby earning for itself an honoured place-second to none-amongst the nations of the world. Its fame spread to lands far and near, even in those remote ages of which no written historical record exists.

Whilst the missionary zeal of our religious emissaries carried the spiritual aspects of our culture to the distant corners of the globe, the spread of our culture on its temporal side was entirely due to our traders and exporters who were our cultural ambassadors abroad. Through the export of its famous textiles the exquisite beauty and fineness of which was almost legendary, ancient and medieval India attained an unrivalled position in the international textile trade, which she maintained till not-so-very-Iong ago. According to George Birdwood, India "was probably the first of all countries that perfected weaving and the art of its gold brocades and filmy smuslins". Can it be wondered then that the textile products art the Indian carftsmen were eagerly 'and even anxiously looked *for* and awaited in the markets of the world? How could competitors ever hope to catch up with a country that was thousands of years ahead of them in this art? They could only ban the use of the. Indian fabrics within their own country. The Roman Senate was probably the first to pass a law (about 1600 A.D.)

prohibiting the wearing of the Indian silk garments by men. A little more than a century later England prohibited by law the use of the Indian calico. This was the beginning of the two-centuries-long unremitting war against India's supremacy in textiles.

As is only natural, a crop of legends has collected around the fineness and the delicacy of the fabrics which the Indian weavers produced with the aid of little more than bare human hands. One such, which is widely believed to be authentic, might be related here. The austere Mughal Emperor Aurangzeb chided his gifted daughter Zebunnisan, for not being properly dressed because her body could be seen through. "But I have no less than seven suits on" protested the lady "one on top of another." "Put on so more then" commanded the Emperor" to obscure the transparency.

As in Architecture, Music and other Fine Arts; India has its own distinctive contribution in the production of textiles so far as artistic designs, harmonious and beautiful colour schemes and, of course, the fineness of the fabrics are concerned. Hand spinning and weaving though carried on a small scale and developed slowly as a cottage industry had yet reached a high stage of perfection, even as early as 327 B.C. We catch glimpses of it in the descriptions, sent home by Alexander's soldiers, of the costumes worn by the people of India. Foreign visitors to this country, on their departure, always carried away with them, as priceless treasures, pieces of silks, brocades and muslins. Cargoes, laden with our wondrous fabrics, were invariably welcomed by other countries. This marvellous technique of the craft which has raised it to the status of a 'Fine Art' has developed through hundreds of generations of craftsmen who formed an exclusive class or guild and handed down from father to son, the knowledge gained and the progress made so that each successive generation of craftsmen added something to it and became more and more efficient, producing better and yet better results. Another characteristic of traditional Indian textiles common to all artistic creations of human hands is that, each piece bears the stamp of the individuality of the craftsmen who produced it. Thus, it has a human interest of its own. As the gifted writer Shm. Kamla Dongerkery has said, the traditional textile of India "reveal the background of rich culture, they give artistic shape and form to the ideas and ideals which inspire the lives of the people" and thus provide one of the *most* reliable hallmarks of the cultural development of the people.

The craft was carried on in the villages of some of the important provinces of India, especially around the coastal towns and cities. It was not restricted, as a profession, only to men alone, but women too, in their own leisure hours, helped their men in spinning and weaving and thus, added to the family income. Indeed, in certain provinces like Assam, tradition required women to achieve some degree of skill in the arts of weaving and spinning and even today unmarried girls in Assam are not considered quite eligible for matrimony, unless they have acquired some grounding in this useful art. But the craft flourished mostly as a result of the patronage extended to it by the royalty and aristocracy, which played a vital role in its development. To suit the tastes of the royal and noble families, the standardol production had necessarily to be very high and ever progressive.

Different types and varieties of textiles were produced in different parts of India and thus we hear of Benaras and Sural Brocades, Dacca Muslins, Cashmere Shawls, Gujrat Patolas a Bandhanis, each famous for its design, quality and colour. In the following pages, an attempt has been made to present, in bare outlines, the chief characteristics of some of the outstandin traditional textiles of India. Illustrations of several of these fabrics have been added to their descriptions.

It would perhaps interest our readers to learn that in a number of ancient families are cherished, even today, as valuable heirlooms, some beautiful specimens of our traditional textiles, A good few of these have been presented to some national museums which proclaim to the nation and to the world of Fine Arts "the glory that was Ind".

Muslins. Dacca (now the capital of Eastern Pakistan) was, for centuries, synonymous with the finest muslins the world has ever produced by hand or machine. In the words of Dr. Forbes Watson "with all our machinery and wondrous appliances we have been hitherto unable to produce a fabric which for its fineness and utility can equal the 'Woven air of Dacca'." Indeea, the, Dacca weavers' Imagic hands produced such exquisitely fine I and delicate fabrics that thee poetic names *'Ab-i'rawan'* (Flowing.i water), *'Baft-Hawa'* (Woven-air), *'Shabnam'* (Evening dew) werel justifiably given to them. Exhibits in some of our museums 'Prove even today that a yald's width of the muslin could easily pass through a lady's ring. Then there are legends which have collected around these fabrics. One of them relates that a five-yard piece of the muslin could be packed in a match box. It is, however, an authenticated fact that a 15 yard piece of 36 inches width *'Mulmul Khas'* (Royal muslin) in the reign of the Emperor Jehangir, weighed only 900 grains. Of course *'Mulmul Khas'* was the Dacca muslin.

The value of Dacca muslins is estimated by the number of warp-threads in a given length of the material as compared with its weight. The greater the length and the number of the threads, with comparatively less weight, the higher would be the price. For instance, a yard's width of *'Mulmal Khas'*, the finest muslin, was known to have 1000 to 1800 threads in the warp, and the weaver took five months to weave 10 yards of the fabric. The weaving of these fabrics could only be done during the rai~y season, because for the weaving of such an extremely fine fabric, a humid atmosphere was essential.

Dacca Saris. Upto the beginning of the 19th century, the Dacca muslin saris, one of the most artistic and beautiful specimens of hand-loom textiles, were counted amongst their valuable and cherished possessions by the women of Bengal. Even today, these saris are beautiful in their designs, but alas the art of making fine muslins is no more, and the saris too are not half as fine as they were in the past.

The saris ate generally grey, white or black with blue or black designs. Occasionally, the patterns are woven in with bright coloured cotton, or silver or gold threads. The Dacca muslins with the woven in pattern are known as *'lamdani'* and the typical designs of flowers or figures used by the Dacca weavers are known as *'Jamdani'* pattern. The saris have very bold and large *lamdani* pattern on the *'Anchal' 'Palloos'* (end portion) and the borders. The rest of the sari is generally covered with numerous small *bootties*.. The common motif is the round design bootties, which suggest *Chameli (Jasmine)* flowers and around these are woven the leaves that recalrthose of the sweet smelling *champak*. When the sprays of flowers are spread all over the sari, it is called a lines, the sari is known as *'Terchha'*. But when the floral desigtt I forms a network which covers the entire field, then the pattern is known as *'Jatar'*.

Sometimes in *Jamdani* design, the flowers are clustered to gether-like the settings of jewels, and then the pattern is given the poetic name of *'Panna hazare'* (thousand emeralds). If a running floral pattern covers the whole field, the expression *Phulwar* is used; but if the flowers are large and life-size the *Jamdani* is caused *Toredar.*

The borders and *Palloo* or. *Anchal* (end portion) of saris are generally decorated with distinctive figure designs. The figures chosen represent birds, animals, and human beings. Peacocks or *'mayura'* and herons or *'hansa'* seem to be popular as bird-figures in the designs of Dacca saris. Also some of the motifs indicate the influence of mythological legends, as well as of the local traditions. The designs are commonly accepted as of Persian origin but many of the designs depict incidents from the Hindu mythology also. The most striking feature of the *Jamdani* pattern is the skill of the weaver in being able to depict the conception of actual motion in the figures he weaves in. Many a design presents birds with flapping wings as if they are about to flyaway. The outlines of figures are always bold, straight and invariably geometric in clean cut lines. Also the intervening space is so well balanced with lines and flowers that a most delightful effect is produced by the combination. It may well be imagined that the weaving of such masterpieces called for the highest skill and craftsmanship with almost unlimited patience as each design must have involved months of work. An idea of the skill of the weavers may be formed from the following description of the process of weaving.

## CHANDERI SARIS

The muslins woven in *Chanderi,* a place for themselves because of near Gwalior, have earned a name their fine quality. *Chanderi* saris are mostly cotton with borders and palloos woven in silk or gold threads. Sometimes mixed threads of silk and cotton are used for weaving the fabrics known as *'Garbhreshmi'*. The palloos of these are very artistically ornamented with gold threads while the ground of the sari is checked, with *bootties* in the centre of each check-square. The borders are woven with double threads which produce an effect of two colours, one on each side. The saris are woven in nine-yard lengths and are very much valued by the Maharastrian ladies.

**Baluchar Buttedar**

Baluchar, a small town near Murshidabad, has become a noted and a highly valued name in the handloom textile history of India. The artisans of the locality produced very artistic figured silk saris known as *Baluchar Buffedar.* In *Baluchar Butfedar,* as in Dacca saris, the *Pallo os* were the most elaborately ornamented portions. The field of the remaining portion of the sari was decorated with small *bootties* of some floral design or figure design of birds. The special feature of *Baluchar Buttedar* is that the design used for the ornamentation show a strong influence of Mughal Art, which is, famous fore its portraits. The weaver of Baluchar chose motifs of human figures and the popular *'Toranj'* (also called *'Kalka'* or *'Guldasta')* which is the most popular motif in weaving, embroidery and printing throughout India under its present appellation 'the mango design'. The richly ornamented *palloo* and a portion of the field of the sari. In the design of the *palloo,* the famous ever popular *'Toranj'* are seeri as though these are set in a frame. The border of the frame is again elaborately decorated with pictorial representation of a lady smelling a flower and stated in a sort of a nich. The interspaces are filled with neatly arranged rows of *'Toranj'* lined with an outer border of flowering plant. The border design which is a simple and straight combination of a small *toranj* and flowering plant is continued for the border of the whole sari.

The subjects for portraits were either a lady or a nobleman dressed in Persian dress and holding a flower or riding a horse or smoking. Though the subjects were always of Islamic origin, yet the *Baluchar Buttedar* were very popular amongst the Hindu ladies. This gesture indicates a complete absence of intolerance existing between the two communities.

Later on, probably due to the desire for the patronage of the English bosses of the East India Company the woven-in-pictorial subjects came to include figures dressed in European clothes, and holding instead of.the traditional flower a wine glass.

The wonderful art of weaving figured fabrics in Baluchar is lost for ever and a few extinct scattered specimen in some muse. urns are the sad mementoes of the perfection it had achieved.

Kam Khwab, Barta and Ab.i-rawan (Brocade). *Kam Khwab* is the name given to real gold brocades. *'Brocades'* is the expression used by Westerners who at first caued it *'Kin Kab'* or *'Cin Cob'*. According to their conception, brocades are thick textiles with woven in pattern very prominently thrown up on the surface of the face of the cloth. Consequently, not only *Kalil Kawabs,* but other similar textiles also like *Buftas, Amrus,* and *Himrus* are known to them by the comprehensive English name *'brocades'*. The poetic name *Kam Khwab* expresses the dream-like beauty and richness of the fabric. It literally means only a little less *(Kam)* than a dream *(Khwab)]or* a dream *(Khwab)* reduced *(Kam)* to reality.

The real *Kam Khwabs* are woven with pure gold threads and the silk yarn is added to provide a bJdy and as a means for colour illuminations. Silver or gold-plated silver threads are also used for keeping down the price but gold is usual. These are heavy fabrics and are generally used for making *palloos,* blouses and for men's halfsleeve Indian jackets, long-coats *(Angarkhas* or *Achkans),* ceremonial robes *(chogas)* and also later on came to be used for curtains and for upholstering the furniture of Public Rooms and Durbar Halls of Princes.

**Bafta or Pot Thans**

These are the brocades in which the major portion of the fabric is in closely woven silk with patterns in gold or silver at regular intervals. This fabric is usually woven in a narrow width of 20" to 30". This is not as heavy or thick as *Kam Khwabs* but is heavier than the other textiles. *Baftas* are also used for blouses and Indian skirts *(Lehngas* or *Damans)* and for men's *Angarkhas* or *Ackhans* too.

Ab-i-rawans. These are silk gauge materials with gold or silver patterns only on certain portions. Silk gauze series with woven-in gold borders and *palloos,* are also called *Ab-i-rawans.* The literal meaning of this poetic name is *'flowing water'*.

Although brocades are a speciality of Banaras for which it famous yet these are also manufactured at Surat and at Ahmedabad shows an *Angarkha* made in Ahmeqabad brocade. The design in the illustration depict the traditibnal floral motif. In old brocades motifs representing animals and human figures enclosed within floral borders were in use.

**Himrus and Amrus**

*Himrus* are the famous silk brocades of Hyderabad (Deccan). The State's second largest town-Aurangabad is the chief centre of the art of Himru-weaving. *Himru* probably a derivative of the Sanskrit *Him* (snow) is a fabric used in winter. The ground is cotton, and silk is used for the brocade on the surface.

The yarn used for weaving *Himrus* is spun so as to produce, when woven, the effect of a warm soft material like wool. The peculiarity of the *Himru* is that the silk thread which is used to form a pattern on the surface of the cloth is carried to the reverse side of the cloth and is collected there *in* clumsy long loops. This forms a rather loose but soft warm layer. Further, the accumulation of the loose threads on the reverse of the cloth, necessitates a lining to all garments made of *Himru* cloth. Thus *Himru* garments make very warm clothing suitable for the cold season.

When silk thread is used exclusively for weaving *Himru,* the fabric is called *'Amru'. Amrus* are generally made in Ahmedabad, Surat and Banaras. *Himrus* are used for men's *Achkans, Chogas,* and for female wear also, *e.g.,* for blouses and *Lehngas.* For generations, the Nawabs of Surat used a special quality of *Himru* fabrics for their dresses which was called the *'Nawab's Himru'.* These fabrics are also used for upholstery and curtains.

Paithani' and Pitambar. *Paithanis* are the beautiful and rich saris made at Pattan or Paithan in the state of Hyderabad (Deccan). These are exquisitely fine fabrics with gauze like texture ornamented with gold patterns woven *in* the texture of the cloth. The borders and the *palloos* which are woven separately as gold brocades are sewn on to the sari. The colour of the sari is usually dark orange, red or yellow, with gold lines arranged *in* a check or in stripes. The inter-spaces are usually filled in with a figure design depicting a goose with an olive branch in its beak.

The borders and the *palloos* have very striking designs in bright and showy colours such as moss-green, canary yellow, and bright pink. The common motif of the designs is the peacock supporting a big vase with sprays of bril1iantly coloured flowers so arranged as to form a Persian cone pattern. The il1ustration of a *Paithani pal/oo* does not have the favourite peacock in the design, but it depicts the harmonious arrangement of the sprays and the surrounding floral design. The vase with sprays is placed between two pillars joined with the *toran* (arch). The design is worked in silks of blue, red, and white colours on a field of pure translucent gold. The whole effect is gorgeous and is eminently artistic in its perfect harmony.

In olden days *Paithanis* were usually woven to order for the Royal family and the weaver took months to complete a single piece. The value of a genuine real *Paithani* ranged between Rs. 2000 to Rs. 3000. In modern times, however, such highly valued and gorgeous *paithanis* are not woven.

*Pitambars* are bright coloured silks 5 yards in length with gold borders sewn on them. These are worn by men specially when performing any of the religious rituals.

Patola is an artistica1ly ortJamented fabric. It is a specimen of wonderful combination of the craft of tie-dying *(Bandhana)* and weaving. *Potola* is mostly in use as a wedding sari in Kathiawar and Gujarat.) In Java and in Indonesia too the *Potola* fabric is used for wedding dresses. The fabric is so exquisitely and so highly valued tbat it is handed down from generation to generation in the family. Women of Gujarat and Kathiawar sure the possession with pardonable pride.

*Potola,* unlike the other ornamented fabrics, is invariably woven in just the plain weave. The elaborate and intricate patterns which mark the *Patola* saris are produced by the wonderful art of *Bandhana* or tie-dyeing. The silkyarn with which *Patolas* are woven, is first dyed by the *Bandhana* process before it is put on the loom. The yarns, both warp and weft, are dyed in the lightest of colours. Then they are stretched on the ground, and the dyer proceeds to mark certain portions to indicate the lines of the desired design. His wife who helps him" in his work, then ties up the marked portions with cotton thread so tightly that the next dye cannot penetrate through to the tied portions. The yarn is then immersed in dye-baths of the desired colours and shades. The operation of tie-and-dye is repeated several times until all the colours and shades required for the planned design have been applied to the yarn. The dyer begins with a light colour, passes next to a bright one and applies the dark colour at the very last.

Then the weaver starts on his job. The warp threads are arranged in pre-arranged sequence, and the weft is then interlaced one by one very carefully to form the planned design. The process of producing a *Patola* is, therefore, a very laborious one and is extremely complicated too. Meticulous care and a good deal of creatiye imagination is needed for marking the correct portions on yarns for dyeing in different colours of the pattern. A very retentive memory is another essential requisite for registering and recalling accurately the sequence of the coloured threads in the pattern. Thus only a few traditional designs are used for *Patola* patterns. The following eight are used by the weavers of Pattan as described by Mr. G.V. Patel:-

"1. *Nari-Junjar bhat* or dancing girl and an elephant design. It has neceslirily a parrot included in it.
2. *Pan bhat* or leaf design. It is said to be the leaf of the sacred *pipal* tree. (Ficus Religiosa).
3. *Rattan chawak bhat* or the cross of diamonds design. It has interspersed diamonds also.
4. *Okhar bhat* or water creat design. The real name of this design on investigation at Pattern, appears to be *akhrot bhat, i.e.*, the walnut design.
5. *PhultJadi (hat* or floral design ) is ganerally enclosed in diapers outlined by a single line. Each diaper contains three flowers.
6. *Wagh-Kunjar bhat* or tiger elephant design. The animals alternate with each other in the design.
7. *Chabri bhat* or basket design. Here each enclosure containing an elephant is made up of four quadrants which look as if forming a basket when two of them are taken together.
8. *Chowkhadi bhat* or a diaper with a double outline design. Each disper includes three flowers borne on a stem."

There is one more design which is used for *dhoties* (the loin cloth worn by men). This design consists of the *devnagri* alphabet and the forms of the letters follow those of the *mantra.s* (hymns) in religious book.

Pattan, a place in Kathiawar, is reputed to be the Lbirth-place of *Patola.* The weavers of Pattan later migrated to Bombay, Ahmedabad and Surat and the making of *Patolas* started at these places also.

Orissa weavers also have adopted the *Patola* technique for weaving their special fabrics like curtains, bed spreads, *odhnis* (scarfs worn over the head and draped round the shoulders and waist by women) and saris. The famous Sambalpore saris are woven like *Patolas.*

Jandhanis. *Bandhan;s* or *Choonaris* are the colourful saris and *odhnis* dyed by tie-and-dye process. These are popular amongst;the women of Gujarat, Kathiawar. Rajputana and Sindh. Premlata Jayakar in her article on 'Tie Dyed Fabrics of India in "Marg" refers to *Bandhanis* in the following words:-

'It is an auspicious garment. A symbol of youth and romance, love play and the *'Sohag'* (wifehood) of Hindu women. It is a garment of laughter'. Indian women are known for their love for bright colours.

Also the tradition and the customs of wearing special colours on different festivals, makes it necessary for them to become familiar with the art of dyeing at home. Thus besides the expert professional dyers almost every Indian girl learns by practice a good deal of the art of dyeing and *Bandhani* work.

*Bandhanis* differ from *Patola* as regards the stage at which they are dyed. Like *Patolas* they are dyed by the tie-and-dye process, which, however, is done after the fabric is woven. The fabric is folded over several times until reduced to a small thick square or a rectangular piece. The piece is then damped and pressed on a block on which a design has been carved. The impressed portions are picked up by the finger nails (the nails are allowed to grow specially for the purpose and are used as a sort of pincers) and are then tied up with cotton thread in a thickness sufficient to resist the dye. It needs training and great skill to pick up all the layers at once and make it crinkle in a particular given manner.

The *Bhalldhanari* or the woman who does the tieing up work works swiftly and ties up all the impressed portions without cutting the' thread but carries it over from one point to the next. The dyeing process is carried out in the same order as in *Patolas,* starting with the light colours and finishing with the dark ones.

But each time, before a new shade or colour is applied the timing up process has got to be repeated usually, the designs used are copies of udarl worn by a Gujrati girl. Few traditional ones and yet practice 0 tieing-up the same design over and over again the *Bandhanaris* become expert to such an extent that they are able to dispense with the process of impressing the fabric with the design. The motif of the traditional designs used for *Bhandhams* reopresents animals, birds, flowers and dancing dolls. When elaborate designs are used the *Bandhanis* are known as *'Gha.' chola'*. In some of the expensive *'Gharchola'* gold threads are woven in to form checks or squares, and then the designs are formed in each of the squares by the tie-and-dye process. The *'Choonaries'* are very light fabrics, and the designs for these consist of dots or pin heads irregularly spread all over the field of the cloth.. Sometimes the dots are grouped together to form a design, and the design is known as *'Ek bundi'* (one dot), *'Char bundi'* (four dots) and *'Sat bundi'* (seven dots).

It might interest our readers to know that in some parts of Rajputana, *e.g.*, Alwar, professional dyers existed till a couple of decades ago, who could dye even the finest muslin in two different colours, one on each side of the fabric at the modest charge of only annas four a yard. This art too is now extinct but specimens can be found in some museums.

Kabnendar or Kalarndar. This is the name given to the hand painted cotton fabrics. They are so called because the artist works out the designs on the material with a fine steel brush not unlike a pen *(Kalam)*. The process is very much the same as used for *Batik work*. The basic principle, namely, resist-dyeing, being common to both. The material is first dyed in pale tpink and then stretched out tight. The artist then traces the outline of the design with his *Kalam* or fine steel brush dipped in melted wax. The fabric is then dyed deep red and finally washed in hot water to melt away the wax. This produces the design in deep red on the background of pale pink.

*Kalasdar* fabrics are also called *Palampores* in the textile trade. They are available in rectangular pieces and are popular with Hindus and Muslims alike. The former use them as canopies for the images of their gods and the latter as praying carpets. Those designed for Hindus portray scenes from the Hindu mythology, whilst those intended for Muslims are of Islamic origin. The latter depicts the conventional *Mihab* with panels forming a frame enclosing the 'Persian Tree of Life' complete with birds and the branches and animals resting under its shade. The craftsmanship and the skill of the dyer is amply borne out by the excellent portrayal of the minutest details with amazing accuracy. The French traveller Bernier who visited India in 1663, during the reign of Emperor Shahjehan, thus describes the fabric which formed the drapings of the Imperial courtyard:

And lined within with those *chittes,* or cloth painted by a pencil of Masulipatam, purposely wrought and contrived with Palampore such vivid colours and flowers, so naturally drawn, of a hundred several fashions and shapes, that one would have said it was a hanging parterre."

Embroidered Fabrics. India is rich in embroidered fabrics. The Indian art of embroidery is an ancient one. Its origin can be traced right back to the Vedic age and it appears to have been well-developed by the time of the great epics. Later on, the influence of incoming races and tribes particularly those which migrated from the hills contributed a great deal toward the development of the art along lines which have given it, its present form. "The stitches employed and art conceptions displayed" by these early artists indicate the extent of their knowledge of the art. It may be noted that throughout the mountains and valleys of India, the art is very popular and is assiduously pursued by men and women. The very colourful embroidery produced is not always intended only for the market, but for home use as well.

A few of the better-known embroidered fabrics may now be briefly described.

*The Punjab Phulkari-Phulkari* really means flower *(Phuf)* work *(kari).* These were the conventional ceremonial shawls worn by the Hindu bride at her wedding when going round the sacred fire with the bridegroom. It is a popular saying in Punjab that when a girl is born in a family the mother, or may be the grandmother, starts embroidering a *Phulkari* to be presented 'to the girl on her wedding day. The women of the *Jat* community are specialised in *Phulkari* work. In almost all the districts of the Punjab, wherever this community settled down the *Phulkari* work Qriginated and has flourished. The peasant women of Rohtak, Hissar, Gurgaon and Karnal are known for embroidering the best *Phulkaries.* Rohtak is considered to be the home of *Phulkari* work.

The stitch for *Phulkari* embroidery is the simple darn stitch. The material on which the embroidery is worked is rough *Khaddar,* but the thread used for the embroidery is silk and the colours used are white, red, yellow or; green". The colour of the *Khaddar* material is invariably red, maroon or brown.

The *Phulkari* motifs generally ale of floral patterns, but geometric patterns are also used. For instance in the *Bagh* or *Shalimar* design, the entire material is covered with a geometric pattern. In orthodox *Phulkari,* however, the same floral pattern is embroidered only at intervals on the cloth, intervening portions of which are left plain. Sometimes, only the borders" are embroidered to a width of 3 or 4 inches and the centre of the material is left plain. This pattern is known as *chobes.*

The choice of the design is wide, but very often they appear to be almost identical because of the close and compact stitches 'used. It is remarkable that the illiterate village women choose designs and the colour schemes which have a charm and beauty of their own and which are worked out unerringly by memory.

*Shishedar Phulkaries.* These are different from the other *phulkaries.* The special feature is that the embroidery is elbellished with tiny mica or looking-glass *(Shisha)* discs fixed to the cloth by button-hole stitches, The material embroidered in this way is very often silk, or even satin. Sindh specialises in this kind of *phulkaries* and is reputed to be its home.

The designs used indicate the influence of the Punjab as well as Cutch in the type of stitches used and the colours selected. The *phulkari* work of Upper

Sindh seems to follow Punjabi *Phulkari* while that of the Lower Sindh bears a resemblance to the Cutch work. Besides the, button-hole stitch, the darning stitch of Cutch are also used in the *Shishedar hulkades.*

*Cutch Phulkari.* This is a name given to the embroidered silk or satin material used for skirts in Cutch. llustration' of the Cutch *Phulkari.* The lower portion of the material is embroidered to form a border, and the rest of the material 'is filled with floral or figure designs. The popular figures chosen by the Cutch artist for embroidery are elephants, peacocks and parrots. Most of the embroidery is done in chain stitch, but occasionally herring bone stitch is also used for finishing the edge of the border.

Chamba Roomals The *Roomals* of Chamba, a state in the Himalayan range, are remarkable pieces of embroidery. Princesses as well as shepherdesses are equally agent in the art which they have adapted as a pleasant or profitable pastime for their leisure hours. The *Roomal* is a square piece of cotton material of the size of a teapoy cover.

The material generally selected is very fine and of delicate texture. The embruidery is worked in double satin stitch which produces the design on both sides of the material and so a *Chamba Roomal* does not have a right or a wrong side.

The motifs for designs are figures, flowers, leaves, forming a sort of frame to set off the central representation of a scene from a mythological story ora legend. It is said of that embroidery that it "imitates the Pahari School of painting". The unique characteristic of the embroidery is that it gives a vivid impression of the embroidered figure being in action or in motion, thereby enhancing the artistic value of the *Roomals.*

Kanthas of Bengat *Kanthas* are not original fabrics, but are made with lengths of old and practically used up cloth. Several pieces of about the same length are placed one on top of another and the edges of all the pieces are sewn together so as to form a padded or quilted rectangular piece. Then the artist proceeds to depict beautiful patterns or scenes from stories from the epics or well known legends, by means of embroidery done in simple running stitch. The work is very fine and neat and accurately executed entirely from memory without the help of any tracing or drawing or any written notes. The *Kanthas* are therefore classed as works of art and one has only to look at them to agree.

The women of Bengal often devote their leisure hours in working *kanthas,* creating beautiful and artistic fabrics out of worn out clothes. Embroidered Fabrics of Kashmir. Embroidered fabrics of Kashmir are world famous. The Pashminn shawls, the silk saris, the Namdas and the various other silk and woollen articles are praised as works of art. The Kashmir embroidery is known as *Kaseeda,* and the stitches used are the satin stitch, the stem stitch and the loop stitch. The darning stitch is also used. The nerring bone stitch is used for the edges of the finished pieces.

The craftsmen, in Kashmir, are generally men and are in variably assisted by boys, often of a tender age, who do the actual embroidery. It is interesting to observe the young boys at work. The master craftsman calls from the design before him, the kind and the number of stitches to be put in: As the instructions are called out the boys work swiftly and deftly with needles and the stitches are completed almost as soon as the master has finished calling them out. The scene resembles that of a small, class in a school with the boys taking down the dictation of the teacher.

Nature's bounteous gifts appear to have been literally showered on Kashmir and it is considered by many to be the nearest approach to "a paradis on earth". Kasheedaartists, therefore, under the inspiration of the beauty of the natural surroundings which they have succeeded in reproducing in their embroidery with such amazing skill, that in the, Words of Shm. Kamla Dongerkerry, "competes with the wealth of Nature's charms".' The Kashmir artist has an inexhaustible treasure of, motifs in: gorgeous colours, ready at hand from which he can draw his designs. In Kashmir embroidery, therefore, we find bunches of fruits, foliage and birds of brilliant hues and-in-numerable shades of colour lend themselves' as charming compounds of varied designs. And the *Turanj* (Of. the mango) is not ignored either. Indeed, it finds a place in almost all the *Kasheedas* of Kashmir The outstanding characteristics of Kashmir embroidery is its elegance and the harmonious grouping of the brilliant colours in the design which produces a restful and soothing effect.

The 'workman-hip is so neat and the stitches so fine that one wonders how bare human hands and that too' of young boys could produce such work. It must, however, be remembered that the training of generations (the art is handed down' from father to son in each of the families) has equipped the artisans with almost an intuilive aptitude for the work which has become second nature to them.

On the other and, the art lives only so long as the line of the craftsmen continue and is likely to become extinct with the line. This, alas, is what started to happen some' time ago due mainly to the rage for replacing human hands by lifeless machinery. Paradoxical as it might appear science is fast stifling and killing art by restricting the scope of arts, creation only to such articles as lend themselves to mass production by mechanised processes.

The modern products of Kashmir are no longer as fine and delicate as in the past, and tend to lose their natural charm. The embroidery is best done in silk but the identical patterns are also repeated in wool. Besides the *Pashmina* shawls arid the silk saris, the industry produces many articles for use in the home.

*Namdas* are the embroidered rugs. The rug is a fluted thick fabric manufactured by the process of pressing wool and cotton together. They are then embroidered in thick wool in bright colour designs similar to those on

*Pashmina.* Kashmir Silk embroidered fabrics besides bring colourful and rich in design and comparatively inexpensive, are within the means of middle dass people. Thus! wider patronage supports the industry and the art has nourished not by the patronage of the great and the rich alone.

Chikankari Embroidered Muslins. The white embroidery on white cotton specially on muslins is known as *Chikan* work.. *Chikankari* is an industry nurtured and developed in the region watered by Ganga and her sister Yarnuna (Jamna). Lucknow in particular is the cradle and unrivalled centre of the art.

*Chikan* work lack in the attractiveness associated with colour yet has a fascination of its own, unequailed by few and surpassed by none. Whilst the *kaseeda* of Kashmir may be rich in colour, reflecting in silk the ravishing beauty with which Nature has gifted the valley; the virgin white *chikankari* is perhaps a translation in shnple cotton of, the purity of the waters of the' sister rivers born in the home of perpetual snows the Himalayas. Much credit for the high position of the art of the Kashmir embroidery goes to its attractive colours. 'The fact that without that colourful aid, the *chikankari* has attained the same eminence, is eloquent of its great artistic value.

Daintiness and delicacy (which are the hall-marks of all Lucknow works of' art) added to a finish and a richness of its own, are 'the outstanding characteristics of *chiktmkari.* The work is sometimes so fine that to the naked eye it presents, the appearance of having been woven-in, along with the fabrics.

There are two main styles of *chikan* embrqidery:

- The flat—this group includes the *bukhi* and the *katao* styles.
- The knotted or the embossed, of which the *murri* and the *phanda* are the well-known varieties. To these may be added a third style, namely, the *jali or* netting which is akin' to the drawn thread work, but is produced in an entirely different 'way.' The drawing out of threads, is, regarded slovenly by the jali embroiderers. 'Instead, the fabric 'is pieced with holes of the requisite size to suit the, pattern and these are then tied up to produce an appearance of net. *Jalis,* are of different kinds variously named *Madras loli, Calcutta Jali,* etc. and are very elaborate and intricate.

The varieties of designs are said to be thirty in number. Considering the fact that the colours are not used to vary the designs, and only the forms and motifs alone produce the various patterns, the number is creditable. The design represents familiar objects connected with daily life, often' grains like rice and millet, in a variety of combinations.

The *bukhia* is the most intricate as it is the most remarkable of *chikandesigns.* It is supposed to be the true *chikan.* Shrimati Kamla S. Dongerkerry considers its comparable to the shadow work of the present day" and thus describes the technique of its production:

"The stitches in this design cover the back of the cloth in the style of the herring-bone stitch, producing an opaque effect on the surface of the fine white fabric and at the same time an outline of motifs of flowers and leaves' with minute stitches resembling the strokes of the back stitch.

'The *katao* produces 'an effect similar to *bukhia.* The save fabric is used to produce an opaque effect and for the outline' a simple prdinary stitch is used.

The *murri* and the *phanda* are used mostly in the patterns representing grains. *Chikankari* work is mostly used for sari borders, blouse or kurta collars and the recent times it was also used for the neck pieces the cuffs, and even the hems 'of men's *angrakhas* (long, coats). It has lately come into use as table linen, tea-cosy covers, and numerous other washable articles of domestic' use. ' Probably due to' the influence or' the Western customers who invariably look for colour in Eastern products, the embroiderers have begun to use coloured threads to a little extent which alas deteriorates the pristice elegance one spotless white.

# 2

# Vegetables Fibres

## INTRODUCTION

### ABACÁ

Abaca is a relative of the banana plant that grows in hot, humid climates. One of the main producers of abaca is the Philippines, which is why the plant has been mistakenly called Manila hemp. Unlike banana plants, however, which are valued for their fruit, abaca's primary benefit is in its long leaf sheaths, which grow to be 12 to 20 feet high. All parts of the stalk—from the outer dark layer to the inner most layers—can be extracted, stripped and processed.

Abaca fiber has been used for centuries to make strong, breathable textiles that are comfortable to wear and long lasting. Abaca is popular for clothing, hats, shoes and slippers. In the Philippines there are at least 150 different traditional weaves. Because of the fiber's tensile strength, abaca clothing has been embroidered, hand-painted, dyed and beaded without any loss of luster and shape.

### BAMBOO

Bamboo fiber resembles cotton in its unspun form, a puffball of light, airy fibers. Many companies use extensive bleaching processes to turn bamboo fiber white, although companies producing organic bamboo fabric leave the bamboo fiber unbleached. To make bamboo fiber, bamboo is heavily pulped until it separates into thin component threads of fiber, which can be spun and dyed for weaving into cloth.

Bamboo fabric is very soft and can be worn directly next to the skin. Many people who experience allergic reactions to other natural fibers, such as wool or hemp, do not complain of this issue with bamboo. The fiber is naturally smooth and round without chemical treatment, meaning that there are no sharp spurs to irritate the skin.

Bamboo fabric is favored by companies trying to use sustainable textiles, because the bamboo plant is very quick growing and does not usually require the use of pesticides and herbicides to thrive. As a result, plantations can easily be kept organic and replanted yearly to replenish stocks. The process of making unbleached bamboo fiber is very light on chemicals that could potentially harm the environment.

In textile form, bamboo retains many of the properties it has as a plant. Bamboo is highly water absorbent, able to take up three times its weight in water. In bamboo fabric, this translates to an excellent wicking ability that will pull moisture away from the skin so that it can evaporate. For this reason, clothing made of bamboo fiber is often worn next to the skin.

Bamboo also has many antibacterial qualities, which bamboo fabric is apparently able to retain, even through multiple washings. This helps to reduce bacteria that thrive on clothing and cause unpleasant odors. It can also kill odor causing bacteria that live on human skin, making the wearer and his or her clothing smell more sweet. In addition, bamboo fabric has insulating properties and will keep the wearer cooler in summer and warmer in winter. The versatility of bamboo fabric makes it an excellent choice for clothing designers exploring alternative textiles, and in addition, the fabric is able to take bright dye colours well, drape smoothly, and star in a variety of roles from knit shirts to woven skirts.

## COIR

Coir is a product of the coconut tree, *Cocos nucifera*, and it is sometimes known as coco fiber. The substance is extracted from the hairy husk of coconuts, and used to create a variety of products such as mats, carpets, upholstery stuffing, and brushes. Coir, pronounced KOY-er, is a very coarse, stiff fiber, and it is also extremely resistant to rot and salt water, making it an ideal material for situations in which other fibers would decay.

The word comes from a Malayalam word, *kayar*, which is derived from *kayaru*, which means "to be twisted." The Malayalam language is spoken in Southern India, particularly in the state of Kerala. The language is related to Tamil, Kota, and Tulu, among many other languages spoken in that region. For English speakers interested in trivia, "Malayalam" is among the longest palindromes in the English language, meaning that it reads the same backwards and forwards.

India and Sri Lanka are the two biggest exporters of coir, accounting for most of the world's supply of the substance. To process coir, coconuts are split so that the stiff fibers are accessible. The outer husk is soaked to separate the fibers, which are sorted out into long fibers suitable for use as brush bristles, and shorter fibers which are used to make things like the padding inside inner coil mattresses. After soaking, the fibers are cleaned and sorted into hanks which may later be spun into twine, matted into padding, or used as individual bristles. Coir takes dye well, and many producers dye coir fiber before export.

There are two primary types of coir. The first type, known as white fiber, is extracted from young coconuts. The white fiber is somewhat finer and more pale in colour than coir from mature coconuts. This type of coir is often woven into yarn which is used to make mats, sacking, rope, and twine. In rope manufacturing, white coir fiber can sometimes be cheaper than other rope ingredients. Brown fiber, the other type of coir, is from mature coconuts, and it tends to be more stiff and unyielding. Many Westerners are familiar with coir in the form of stiff doormats, which can be left out in all weathers because of coir's rot resistance. Coir matting can also be found inside some upholstered products, and a close inspection of an inner spring mattress may also yield coir matting, depending on when and where the mattress was made. Coir matting is also sometimes made available in the form of a substrate for plants, since it can be impregnated with water and seeds to sprout flowers and groundcover. The matting will keep weeds back while the plants establish themselves.

## COTTON

Cotton is a natural fiber harvested from the cotton plant. Cotton is one of the oldest fibers under human cultivation, with traces of cotton over 7,000 years old recovered from archaeological sites. Cotton is also one of the most used natural fibers in existence today, with consumers from all classes and nations wearing and using cotton in a variety of applications. Thousands of acres globally are devoted to the production of cotton, whether it be new world cotton, with longer, smoother fibers, or the shorter and coarser old world varieties.

Cotton is in the mallow family and produces delicate, lovely flowers. Other members of the mallow family include hollyhocks and hibiscus, used to brighten gardens all over the world. The cotton fiber forms around the seeds of the cotton plant and is designed to help carry the seeds long distances on the wind so that the plant can distribute itself. Early humans realized that the soft, fluffy fibers might be suitable for textile use and began to breed the plant, selecting for fluffy, easily spun varieties.

After harvesting, cotton must be combed to remove the seeds. This used to be a laborious process until the invention of the cotton gin, which quickly separates the seeds from the fiber and combs them for spinning. While a single cotton fiber is not terribly strong, when multiple curling fibers are straightened and twisted together, they form a strong, smooth thread that can be knitted or woven, as well as dyed.

Cotton is somewhat flammable, especially lighter cottons that hold a lot of air. Some cotton is chemically treated to reduce flammability. Many cottons are also blended with other natural fibers, such as linen, for particular properties, or to add texture and strength to the fiber. Cotton can be woven or knitted. It can also be turned into flannel, corduroy, muslin, and a variety of other fabrics used so universally that the American Cotton Council uses "the fabric of our lives" as a tag line. Cotton also carries environmental controversy, particularly in the developing world, where dangerous pesticides are heavily employed. Cotton is subject to infestation, and therefore many growers heavily douse the plant in pesticides that are harmful to human and animal health, as well as herbicides to eliminate competition for resources. A number of producers also genetically modify the plant, which many outside the industry view as a questionable practice. Cotton also has very large water requirements, which may place stress on nations with limited water resources. In the late 20th century, there was a push for organic, sustainable cotton grown and harvested without the use of pesticides and human exploitation. This cotton is significantly more expensive than conventionally farmed cotton, however, and may not be practical for most consumers.

## FLAX

Flax is a member of the genus *Linum* in the family Linaceae. It is native to the region extending from the eastern Mediterranean to India and was probably first domesticated in the Fertile Crescent. This is called as Agasi/Akshi in Kannada, Jawas/Javas or Alashi in Marathi. Flax was extensively cultivated in ancient Ethiopia and ancient Egypt. In a prehistoric cave in the Republic of Georgia dyed flax fibers have been found that date to 34,000 BC. New Zealand flax is not related to flax, but was named after it as both plants are used to produce fibers.

Flax is an erect annual plant growing to 1.2 m (3 ft 11 in) tall, with slender stems. The leaves are glaucous green, slender lanceolate, 20–40 mm long and 3 mm broad. The flowers are pure pale blue, 15–25 mm diameter, with five petals; they can also be bright red. The fruit is a round, dry capsule 5–9 mm diameter, containing several glossy brown seeds shaped like an apple pip, 4–7 mm long. In addition to referring to the plant itself, the word "flax" may refer to the unspun fibers of the flax plant.

## FLAX FIBERS

Flax fibers are amongst the oldest fiber crops in the world. The use of flax for the production of linen goes back at least to ancient Egyptian times. Dyed flax fibers found in a cave in Dzudzuana (prehistoric Georgia) have been dated to 30,000 years ago. Pictures on tombs and temple walls at Thebes depict flowering flax plants. The use of flax fiber in the manufacturing of cloth in northern Europe dates back to Neolithic times. In North America, flax was introduced by the Puritans. Currently most flax produced in the USA and Canada are seed flax types for the production of linseed oil or flax seeds for human nutrition.

Flax fiber is extracted from the bast or skin of the stem of the flax plant. Flax fiber is soft, lustrous and flexible; bundles of fiber have the appearance of blonde hair, hence the description "flaxen". It is stronger than cotton fiber but less elastic. The best grades are used for linen fabrics such as damasks, lace and sheeting. Coarser grades are used for the manufacturing of twine and rope. Flax fiber is also a raw material for the high-quality paper industry for the use of printed banknotes and rolling paper for cigarettes. Flax mills for spinning flaxen yarn were invented by John Kendrew and Thomas Porthouse of Darlington in 1787.

## HEMP

Hemp is a natural fiber product of the *Cannabis sativa* plant. Astute readers may be aware of other byproducts of this plant, but hemp is produced from a type of *Cannabis sativa* specifically bred to yield long fibers. Cultivation of hemp for industrial purposes has been undertaken for thousands of years, and hemp was used to manufacture rope, canvas, paper, and clothing until alternative textiles for these purposes were discovered.

Traditionally, hemp has been a very coarse fiber, which made it well suited to rope but less than ideal for clothing designed to be worn against delicate human skin. Advances in breeding of the plants and treatment of the fibers have resulted in a much finer, softer fiber, which is ideal for weaving into clothing. While hemp clothing in the late 20th century came to be associated with fringe movements, it was once widely utilized as a textile: the word *canvas*, for example, is related to *Cannabis*, one of the original components of canvas.

In addition to providing useful fibers, hemp seed also has high nutritional value, and the plant can be used to make biodegradable plastics, some fuels, and a variety of other things. While hemp is unlikely to save the world, as many proponents are fond of saying, it is an underutilized vegetable resource. Hemp is rich in healthy fats and some vitamins, depending on how it is grown. As a result, it is frequently used in skin salves and balms, as well as in nutritional supplements.

Hemp clothing tends to be strong, insulating, absorbent, and durable. This durability makes it well suited to garments that will see hard wear, because hemp fibers can last up to three times longer than cotton fibers. Most frequently, hemp clothing is woven, although the fibers tend to form chunkier threads than other natural textile components like cotton. Hemp can also be used in knits.

Untreated hemp fiber is pale blonde in colour and takes dye well. Many hemp textile products are coloured with plant dyes, which gives hemp an undeserved reputation for being dull in colour. In fact, hemp can be dyed as vividly as other textiles like cotton.

## JUTE

Jute is a type of plant fiber used to make common items such as rope, twine, chair coverings, curtains, sacks, hessian cloth, carpets, and even the backing used on linoleum. This is accomplished by spinning the fiber into a coarse thread. Despite the fact that jute tends to be rough in texture, fine threads of it are sometimes used to create imitation silk. In addition, jute is increasingly being looked at as an alternative source for making paper, rather than cutting down trees for pulp.

The thread created from jute is quite strong, yet it is among the cheapest of natural fibers available. It also has exceptional insulating properties, low thermal conductivity, and antistatic characteristics. Nonetheless, synthetic materials are replacing jute in many applications, because they are still less costly to create and more efficient to use. This is partly because jute has a tendency to become brittle and to yellow in sunlight. It also tends to lose its strength when wet and can become infested with microbes when used in humid regions.

There are several applications for which jute is still used instead of synthetic fibers. These applications are mostly limited to those that require the use of a material capable of biodegrading. Pots for plants that are planted directly into the ground with the plant, for example, are often made of jute. Jute cloth is also used in landscaping projects, in order to prevent erosion while still permitting natural vegetation to grow.

Jute is also considered to be a possible alternative to wood. This is because its stem contains a woody inner core. Taking just four to six months to grow to maturity, jute can be harvested much more quickly than trees. Many hope to be able to use jute in order to slow down or prevent deforestation.

The majority of the jute used today is grown in the Ganges delta. This is because the plant prefers climates that are both warm and humid, with temperature ranging from 68 to 104°F (20 to 40°C) and a relative humidity of 70-80%. It also requires about two to three inches (5 to 8 cm) of rainfall per week. China is the next largest producer of jute.

## KAPOK

Kapok is a tropical tree of the order *Malvales* and the family *Malvaceae*, native to Mexico, Central America and the Caribbean, northern South America, and to tropical west Africa. The word is also used for the fibre obtained from its seed pods. The tree is also known as the Java cotton, Java kapok, or ceiba. It is a sacred symbol in Maya mythology.

The tree grows to 60-70 m (200-230 ft) tall and has a very substantial trunk up to 3 m (10 ft) in diameter with buttresses. The trunk and many of the larger branches are often (but not always) crowded with very large, robust simple thorns. The leaves are compound of 5 to 9 leaflets, each up to 20 cm (8 in) and palm like. Adult trees produce several hundred 15 cm (6 in) seed pods. The pods contain seeds surrounded by a fluffy, yellowish fibre that is a mix of lignin and cellulose.

## KENAF

Kenaf, *Hibiscus cannabinus*, is a plant in the Malvaceae family. *Hibiscus cannabinus* is in the genus *Hibiscus* and is probably native to southern Asia, though its exact natural origin is unknown. The name also applies to the fibre obtained from this plant. Kenaf is one of the allied fibres of jute and shows similar characteristics. Other names include Bimli, Ambary, Ambari Hemp, Deccan Hemp, and Bimlipatum Jute. It is an annual or biennial herbaceous plant (rarely a short-lived perennial) growing to 1.5-3.5 m tall with a woody base. The stems are 1–2 cm diameter, often but not always branched. The leaves are 10–15 cm long, variable in shape, with leaves near the base of the stems being deeply lobed with 3-7 lobes, while leaves near the top of the stem are shallowly lobed or unlobed lanceolate. The flowers are 8–15 cm diameter, white, yellow, or purple; when white or yellow, the centre is still dark purple. The fruit is a capsule 2 cm diameter, containing several seeds.

## PIÑA

Piña is a fiber made from the leaves of a pineapple and is commonly used in the Philippines. It is sometimes combined with silk or polyester to create a textile fabric. The end fabric is lightweight, easy to care for and has an elegant appearance similar to linen. Pina is also related to a drink from Puerto Rico known as Pina Colada.

Piña comes from the leaves of the pineapple plant. “Each strand of the hand scraped Piña fiber is knotted one by one to form a continuous filament for hand weaving into the Piña cloth”. The piña fiber is softer, and has a high luster, and is usually white or ivory in colour. It is also called Pina in Spanish.

## RAFFIA PALM

The Raffia palms (*Raphia*) are a genus of twenty species of palms native to tropical regions of Africa, especially Madagascar, with one species (*R. taedigera*) also occurring in Central and South America. They grow up to 16 m tall and are remarkable for their compound pinnate leaves, the longest in the plant kingdom; leaves of *R. regalis* up to 25.11 m long and 3 m wide are known. The plants are either monocarpic, flowering once and then dying after the seeds are mature, or hapaxanthic, with individual stems dying after fruiting but the root system remaining alive and sending up new stems.

## RAMIE

Ramie is a flowering plant which is native to Asia. It is harvested and processed to yield strong fibers, also called ramie, which are used in the production of textiles, twine, upholstery, filters, and sacking. Like flax, jute, and hemp, ramie is considered a bast fiber crop, meaning that the usable portion of the plant is found in its connective tissue structures. The plant is widely cultivated in several Asian nations, which export ramie around the world. The scientific name for the plant is *Boehmeria nivea*, and it is also sometimes called Chinese Grass. The plant grows in the form of stalks with heart shaped leaves which sprout up from an extensive underground root system. Ramie is in the nettle family, and it has the characteristic small silvery hairs associated with nettles, although the hairs do not sting. The ramie stalk can be harvested up to six times each year in favorable cultivating conditions, although three to four crops of ramie annually are much more common.

The plants must be extensively processed to yield ramie fiber. A series of beatings, washings, and chemical treatments extracts the usable part of the plant and de-gums the fiber so that it will be usable. Once processed, ramie can be spun into thread or yarn, and it is sometimes also blended with other textile materials to make it more versatile.

Pure ramie is very strong, resistant to mold and bacteria, lustrous, and it holds its shape very well. However, ramie is also stiff, not terribly elastic, and sometimes difficult to work with because it can be very brittle. In addition, ramie does not take dye very well. All of these shortcomings make ramie more expensive than similar plant fibers, such as linen. Although the material has been used in the production of textiles for thousands of years, many producers prefer to produce it in blends rather than using it plain.

Like linen and other textiles made from woody fiber, ramie requires special care. Ideally, the fabric should be hand washed cold and not wrung or heavily pressed before being laid flat to dry. Once dried, ramie should be stored flat, and it should not be sharply creased or folded. In some cases, ramie can also be dry cleaned or machine washed on a gentle setting, and the care directions on the fabric should always be carefully followed to avoid damaging it.

## SISAL

Sisal (*Agave sisalana*) is an agave that yields a stiff fibre traditionally used in making twine, rope and also dartboards. The term may refer either to the plant or the fibre, depending on context. It is sometimes incorrectly referred to as *sisal hemp* because hemp was for centuries a major source for fibre, so other fibres were sometimes named after it.

The plant's origin is uncertain; while traditionally it was deemed to be a native of Yucatan, there are no records of botanical collections from there. Gentry hypothesized a Chiapas origin, on the strength of traditional local usage. In the 19th century, sisal cultivation spread to Florida, the Caribbean islands and Brazil, as well as to countries in Africa, notably Tanzania and Kenya, and Asia. The first commercial plantings in Brazil were made in the late 1930s and the first sisal firer exports from there were made in 1948. It was not until the 1960s that Brazilian production accelerated and the first of many spinning mills was established. Today Brazil is the major world producer of sisal. There are both positive and negative environmental impacts from sisal growing. Traditionally used for rope and twine, sisal has many uses, including paper, cloth, wall coverings and carpets.

## WOOD FIBRE

Wood fibres are usually cellulosic elements that are extracted from trees, straw, bamboo, cotton seed, hemp, sugarcane and other sources.

The end paper product (paper, paperboard, tissue, cardboard and etc.) dictates the species, or species blend, that is best suited to provide the desirable sheet characteristics, and also dictates the required fibre processing (chemical treatment, heat treatment, mechanical 'brushing' or refining etc.).

In North America, virgin (non-recycled) wood fibre is primarily extracted from hardwood (deciduous) trees and softwood (coniferous) trees. Wood fibres can also be recycled from used paper materials.

Wood fibres are treated by combining them with other additives. They are then processed into a network of wood fibres, which constitutes the sheet of paper.

# COTTON

The origin of cotton can be traced back to India the original home of the best and finest cotton produced over world. The cultivation of cotton gradually spread throughout Asia to few parts of Africa and eventually, to the Southern States of U.S.A., which now grow more cotton than the rest of the world. Cotton is the white, downy fibrous substance covering the seed of the cotton plants.

**Fig.** Cotton Field

The seeds with this covering are encased in pods which grow on the cotton plant and burst open when ripe, disclosing the white, downy covering of the seed now grown into cotton fibres. The cotton plant grown in the tropics requires a climate with six months of summer weather, to blossom and produce the pods. The cotton fibre is the shortest of all the textile fibres. Its length varies from 8/10 of an inch to 2 inches.

Cotton with short length fibres is technically called 'short staple' and the one with long fibres is called 'long staple'. The latter fibre is valued more as it is utilised for making the fine qualities of cloth for which it is specially suitable, as it is easy to spin and produces a strong, smooth yarn. It is also suitable for mereerisation.

### *Process of Collecting of Cotton Pods*

**Fig.** Collecting of Cotton Pods

When the pods burst they are picked from the plants and collected. If left on the plant, the cotton gets discoloured and dirty by explore to the plant, the cotton gets discoloured and dirty by exposure to the sun and weather.

### *Process of Ginning*

The process by which the seeds obtained from the pods are separated from the cotton fibres covering them. This process is carried out by means of machinery, and the seperated fibres have now become 'Cotton'.

### *Practice of Baling*

The cotton is then pressed into bales which are wrapped with jute sacking and bound with steel bands. The cotton is then supplied to the mills in bales.

### *Cotton Preparation before Spinning*

In the mills, the bales are opened and the cotton is pulled out and beaten to remove loose dirt or any other foreign matter loosely present in it. The opened out cotton is then compressed into a sheet called lap.

***Process of Carding***

The lap on a cotton sheet is passed through a machine called a 'card'. In such machine, the cotton is thoroughly cleaned of ail attached dirt and foreign matter and the matted fibres are separated and laid nearly parallel to each other.

***Process of Combing***

This is an improved combination of the carding process. In this, all short fibres are elirr inated and even long fibres are laid more parallel, thus forming a film like sheet of fibres.

Fibres which are carded and combed are of more even long and smoother than those which are only carded.

***Process of Slivering***

At the end of the 'carding and combing' processes, the film-like sheet is drawn into a strand about one inch in diameter called the 'Silver', which is collected through a coiler into a can. At this stage, the cotton is ready for spinning.

## SPECIAL FEATURES OF COTTON

### ITS FORMATIONS

The cotton fibre is composed chiefly of cellulose which constitutes 88-90 per cent, water 5 to 8 per cent and other natural impurities.

***Construction***

When raw, the fibre has a tube-like structure containing sap. When the fibre ripens, the sap dries up, the tube collapses and the fibre becomes like a flat, twisted ribbon. A central canal, the lumen runs through the fibre and in rich cotton it appears like an irregular loop. In dead cotton it is practically absent. Such fibres are weak and brittle. Cotton fibre, therefore, has no lustre or elasticity.

***Fibre Shrinking***

The fibre itself does not shrink but the fabrics: made with it which have been stretched in the finishing processes do.

***Impact of Moisture and Friction Cotton***

It is not affected by moisture or friction and is stronger when moisture content increases. Therefore, it lends itself to the washing process.

***Hygroscopic Moisture***

Cotton does not hold moisture so well as wool or silk, but absorbs it and so feels damp, much more quickly. It also rapidly spreads over throughout the material.

### Good Conductor Heat

Cotton is a better conductor of heat than wool or silk but not as good as rayon.

### Dyes of Cotton

Cotton takes in dyes better than linen but not as readily as silk or wool. If a mordant is utilised, cotton is easy enough to dye. Mordant colours, direct or substantive dyes should be applied to cotton.

### Acids and Alkalies Action

Strong acids will destroy the fibres soon. Dilute inorganic acids wi1l weaken the fibres and if left dry will rot it. Thus, after treatment with acidic solutions, cotton articles should be thoroughly rinsed. However they are very little affected by organic acids. They are also very resistant to alkalies, even strong caustic alkalies at high temperature and pressure. In 18 per cent NaOH cotton fibres swells, spirals, twists uncoil and shrinks and becomes thicker. The resultant fibre is smoother, lustrous, stronger and has increased water and dye absorption.

### Bleaching Impact on Cotton

All bleaches can be safely used on white cotton. Linen is believed to have been utilised nearly 10,000 years ago by the European Neolithic people. Fragments of the cloth have been discovered in parts of Switzerland, the home of the Neolithic people.

In Egypt, the Mummy cloth or the material wrapped round the preserved dead (mummies) thousands of ears ago has been findout to be linen.

Thus, linen has been an important fabric in the past. Even today, a modern house-wife is proud of having lipen for use in her household.

Linen is the bast fibre obtained from the inside of the stems of the flax plant. Flax is grown inmoist, temperate climates. Most of the European countries and Egypt cultivate it.

### Its Requirement

When the stems of the plant turn yellow at the base, and when the seeds turn from green to pale brown, the plants are pulled out by the roots. Pulling out by the roots gives long, unbroken fibres.

### Process of Drying and Rippling

The flax after it has been pulled out is tied up in bundks and left to dry for a few days. The leaves and seeds are then removed from the stems by a process called rippling. This is done by passing the head of the plant through a coarse comb or a machin. Care is taken not to break or injure the stem. After the removal of the leaves and seeds the stems are again tied up in bundles.

## Some Other Small Fibres

### *Hemp Fibres*

This is also a bast fibre stronger than linen, or jute, and is dark brown in colour barring the Manila hemp, which is white. It is not used. For weaving fine cloth but is used for making gauzes, ropes and webbing.

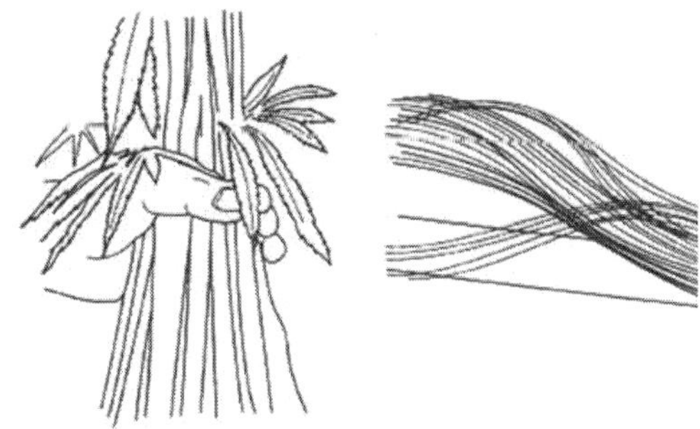

**Fig.** Hemp Fibres

### *Jute Fibres*

**Fig.** Jute Fibres

This is also a bast fibre of plant 'Cochorous Capsularis' that grows in India, to a height of 12 ft. The fibre is prepared in the same way as flax fibre.

These fibres are weaker than linen, are short, lustrous and smooth. Jute is affected by chemical bleaches, and so it is never made pure white. It is utilised for floor covering, gunny bags, binding threads, and also used for dress materials and saris.

### *Kapok*

Fibres is a cotton like fibre. It is get from a tree grown chiefly in Java, West Indies, Central America and India. The fibre is finer than cotton, is silky in look but is not suitable for spinning.

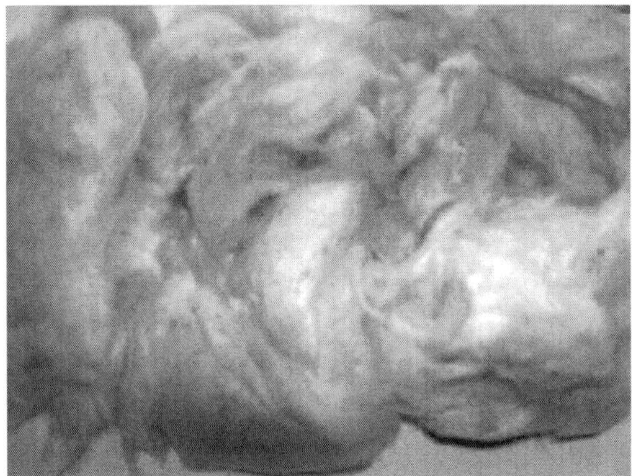

**Fig.** Kapok

It is used only for filling up mattresses and pillows.

***Ramie Fibre***

**Fig.** Ramie Fibre

This is a bast fibre like linen, and is sold as its substitute; It is obtained from a species of nettle plant that grows mainly in Bengal. It also grows in China, Egypt, Java and Japan. The plant grows to a height of 4 to 8 feet. The stalks are cut and water retted, the excessive gum is removed by a chemical. The fibres are combed and straightened.

The long fibres are known as 'line', and the short ones are knownas 'noils'. Linen fibres are spun and woven into a cloth, which is durable and has crispness which makes it roped for attractive table linen. Ramie cloth is known as China or Canton Linen. The noils are used for making canvas *wadding* and cardage.

## MATERIALS AND PROCESSING IN FIBRE PRODUCTS

Quality has two meanings. It is mainly concerned with material and processing that are used in the products. Further it means costly materials or processing or both have been brought into use wish a view to achieve the desired features. High price and quality are generally associated because the particular end-use has stringent needs, which are difficult, and hence expensive, to meet. In view of this relationship, quality in this sense can be referred to as expense quality.

Since the more expensive material and processing tend to give a greater range of properties, this title' has a real meaning. It has, not with standing, certain limitations in that inexpensive materials or processing may produce a product, that is perfectly suited to a special end-use. High or low-expense quality does not imply any judgement on the merits of the product: it relates only to the kind of product. In fact, of course, the objectives is to produce an article that satisfies the demands of the end-use it was designed to serve.

For this cause, it is preferable to, use the term design quality when the merits of a product are judged from the standpoint of end-use suitability. Whether a yarn or fabric has a high end-use suitability relies upon how the design has been translated into reality through production. This introduces the concept of quality measured in the degree of conformity to specification.

Achieving conformance is a production problem and this aspect of quality can usefully be referred to as 'production quality' Circumstances will affect the relative importance of these two aspects. End-use suitability does not merely imply the presence of a particular quality characteristic; rather is it a combination of many separate or related, characteristics or both. In certain cases, once the material has been chosen and the processing sequence decided upon, most of the features are already uniquely determined. In others, the important characteristics may be quite different. In general, however, the characteristics fall into three groups. First, there are those that rely solely on the raw materials and their reactian to acids, heat, etc.

Secondly, there are those qualities that are a reflection of the design and maintenance of the machines utilised and the skill and attension of the operatives attending such machines. For instance, if a roller forming part of the drafting system on a spinning frame is eccentric, the material being processed will contain a periodic wave.

The size and intensity of the wave will rely on the diameter of the roller, the amount of attenuation, and the degree of eccentricity. Finally, there are those that are the combined result of both the material and the processing. For example, in cotton yarn the incidence of neps is mainly dependent on the percentage of immaturc fibres present.

At the same time, the condition of the carding engine, particularly the state of the clothing and the frequency of stripping and grinding, can also influence the number af neps in the finished yarn. As knowledge of materials and machines becomes greater, the interrelationships divulged increase the importance of this third group.

## Meaning

As with quality, there are various opinions as to the meaning of quality control. To some, quality control is only a synonym for testing, whereas others associate it primarily with the application of statistical methods to quality problems. Both views are unnecessarily restrictive.

Broadly, quality control is regarded from two several standpoints; First, it is regarded as the control of all the factors that play a part in determining the quality of the final product. In this sense, it is a basic part of management. Secondly, it is used to describe a series of techniques or procedures that enable management to discharge its quality responsibilities more effectively. To call the latter' statistical quality control' is no solution, since this is concerned only with a few of the many techniques available. Quality control cannot be achieved in a complex textile undertaking without both management and technique. Practically, the deciding factor is that of, management. Effective control depends on managerial policy and action; the techniques, however precise, can serve to improve such action, but they cannot replace it, They spotlight differences between the actual and required standard and, where feasible, measure such differences quantitatively. Control itself depends on the taking of specific action to remedy such differences and this, in turn, involves management.

If the above comments are borne in mind, quality control is defined as being related to all the procedures utilised and action taken to ensure that the quality characteristics of the product conform to the needed standard. This standard may, or may not, be written down, and it may not be especially precise; nor need it cover all the feasibles. The measure of the effectiveness of quality control is then simply the degree to which such conformity is attained.

## Standard's Role

Fundamental in quality control is the concept of a standard of quality performance. Here again the word 'standard' is subject to various interpretations. It is important, however, to consider the many meanings in order to provide a basis for a realistic and consistent quality policy. There are at least three-types or aspects of standard that merit consideration. They all play a significant part in establishing a yardstick against which the effectiveness of quality control can be measured.

## Standards of Product

The first important standard is the commercial one. This is the quality level desired for the market in which the product is to be sold. In the textile industry the practice is varied. Although cloth constructions are generally laid down, there is considerable variation in the amount of detail on performance characteristics.

Specifications of the permissible level of fault incidence likewise vary from the exceptionally detailed to the very vague. For yarnsl it is still common to produce to a specification covering only average count and minimum strengthl. Both these characteristics being judged by a somewhat haphazard series of tests on small samples. It is worth noting, however, that in many sections of the industry the trend is towards more detailed specifications, embodying not only more characteristics but also more systematically defined permissible limits. There are many factors influencing this trend but only a few requirements be mentioned here. First there is the growing importance and influence of integrated manufacturing units.

Controlling as they do part, or all of their processing they admire how various levels of yarn or cloth quality are achieved and how they affect performance at subsequent processing. On those occasions when they supplement their supplies from outside sources, they tend naturally to lay down specifications as similarly detailed as those used within their own units.

Secondly, there is the increasingly important role of the large institutional buyers. Several of these organizations have a direct financial interest in some of the most modern spinning, weaving, and knitting units. Besides, their influence is increased, since they are also the largest customers of some other units.

Laying down detailed quality requirements for the garments they sell, they force the firms at earlier processes to impose similarly stringent demands on their suppliers. Thirdly, one of the factors that has hitherto prevented the preparation of standards on several features is that of inadequate instrumentation. Developments in recent years have enabled more characteristics than previously to be accurately measured and compared with reproducible standards.

Even so, the influence of these forces should not be exaggerated. Many textile firms, in several sections, are still working in a situation where the standards in use are very loosely defined.

Again, where standards exist, they still fluctuate as cyclical trade fluctuations affect the relative bargaining strength of buyers and sellers. This situation has had a fundamental influence on both the level and the control of production quality. Although there are often no clearly laid-down commercial standards, this should not be regarded as a reason for reducing attempts to control quality.

When certain customers are not able to measure quality features in a systematic and precise way, it is incorrect to assume that they do not regard them as important, even though this is not clearly stated. Secondly, several of the quality characteristics are an index not only of the suitability of the product for later use but also of the ease of processing in the producing mill itself. For example, it may be worth while for a spinner to produce a yarn stronger than a particular weaver requires because such a yarn can be processed with fewer breakages at the spinning frame.

With reference to many of the characteristics not closely determined by the raw material used, it is often expensive to process low-quality product. Better product quality is often associated with better and cheaper processing at the producer's mill as well as' at that of the purchaser. Without adequate control procedures and checks, however, it is not feasible for a manufacturer to evaluate the gains made in his own processing. Briefly, a producer is not able to make a consistent decision on what quality to aim for if account is taken only of the demands made by his customers.

**Quality Level**

Commercial standards relate to the quality level desired for the particular market in which the product is to be sold. More simply, they are those that are referred to by the buyer and seller in relation to their commercial transactions.

Whatever commercial arrangements are made, however, the individual producer should aim at laying down performance standards at each processing stage in as precise and detailed a form as possible. These constitute a second group, which can be called mill quality standards.

For example, to evaluate the setting, scouring, and maintenance procedures at the doorframe, it is essential to measure the continuity of the sliver produced, with regard both to its overall level and to the incidence of periodicities. This could in no way be called a commercial standard, in that no one purchasing the yarn would be directly related to its performance at an early stage of processing.

The objective of mill quality standards is to provide management with a measure of the mill's quality performance. When standards for various quality characteristics are laid down and agreed with the personnel liable for their attainment, management is in a strong position to locate weaknesses in staff, materials, or machines and to take proper action when and where standards are not met. Without objective quality indices, management cannot assess quality performance and its positton is correspondingly weakened. The setting of these standards, which must be realistic, is not easy.

What is realistic in any special situation depends on the inherent variation in the process and on what is possible, bearing in mind the quality of the material or labour employed. To get this information very often requires complex process-capability studies and, as so often, the greatest enemy is wishful thinking.

If the customer is demanding a standard higher than can be met with existing machinery or materials, no useful objective is to be served by arbitrarily tightening mill standards. Commercial and mill standards tend to coincide, in the long run. This is because, in general terms, if the quality is lower that required, the product will not be sold, and if the quality is higher than that desired, it will adversely affect the firm's competitive position because the product will cost more.

In general terms is the key phrase here as the here, because the determination of the standards is taking place in a changing situation, and the actual position sometimes, if ever, reaches equilibrium.

Again, as has been previously stated, it may still pay in certain cases to produce a quality higher than is required if the processing economies that result justify it. The discussion has manufacturing one quality of implicitly assumed that the producer is manufacturing one product. In actual practice, this is rarely so, and often there will be many different quality levels in a mill. It will not always be feasible or desirable to achieve coincident commercial and mill-quality standards, on all of them.

The quality of material and the length and type of processing can be varied, but the mill standards for operative performance and machine condition must be geared to the highest quality being produced. In these respects, long-term quality standards must be the dominant ones, for it is notoriously difficult to restore lowered quality.

Several firms found that they paid dearly for condoning poor operative methods and machine maintenance in the period of lowered commercial standards immediately after the second world war.

## TECHNOLOGICAL AND PROCESSING STANDARDS

Eventually, mention should be made of the role of general technological and processing standards. These 'norms' which are Published in different technical journals, as well as in machinery manufacturer specifications, are useful to management despite their invitable generality. They generally relate to what is feasible in the present state of technical knowledge and machine development.

To illustrate, one may quote some of the performance features to be hoped from a modern speed frame employed on the cotton system for coarse-medium counts. The coefficient of variation of weight per unit length within a bobbin should be less than 1 per cent; the Uster regularity percentage should not cross 3.5 per cent and there should be no periodicities; it should be possible to change the staple length processed by 3/16 in.

Without changing settings, with a maximum deterioration of 0.5 per cent. This information is useful in helping management to evaluate the feasible gain to be made in quality in relation to the capital expenditure necessary to achieve it. Besides, it gives an indication of the hoped future level of competition in quality.

Unfortunately however, it does not provide much assistance to a Particular firm in fixing its mill-quality standards. Most firms are more interested in finding out how they compare. With other firms utilising similar processing and machines of comparable age, rather than with some theoretical ideal.

Such information is rarely available, and there is a tendency for managements to attribute all the quality differences to better machinery or various processing. Actual inter-firm comparison are so infrequent, and information is so scanty, that it is difficult to know if mill-quality standards are realistic. There is undoubtedly a great deal of confusion over standards in many textile. firms.

This generally results in an unwillingness or inability to lay down what is to be achieved at the many stages of manufacture. It is important that the scope and aim in any particular case should be laid down and clearly understood by all concerned. Only then is it feasible to devise a quality policy to meet these aims and, by measuring the results of such a policy, to evaluate its effectiveness.

There are many difficulties in drawing up the many standards that form an inevitable part of a quality policy. One approach, which the author has found to be of value in widely differing circumstances, is that of 'quality audit'. It is not possible here to give a detailed description, and a brief outline only is included below. A quality audit can be regarded as a logical application of the scientific method to the problem, of qua!ity improvement.

In the beginning, information is collected on all those factors that affect the final quality of the product. The central feature of this initial stage is the series of observations carried out by a team drawn from the mill staff. These obseryations give details of the processing conditions on the various machines and the methods employed by the operatives attending those machines. Concurrently, inspections are carried out to assess the mechanical state of the machines in question and to measure sure the environmental conditions in which processing is being undertaken.

To complete the picture, the material itself is tested at various stages of processing. This provides more information on the effects of these various factors on the quality of the final product. These observations, inspections, and tests are planned in such a way that it is feasible to interpret and analyse the, interrelationships that they reveal by the use of nothing more than the simplest statistical methods. This analysis forms the second stage of the audit, the object being to trace back. important quality differences to the relevant technical and operating conditions.

From the results of the analysis a report is drawn up, this providing a complete account of the quality position of the mill. While the various kinds of remedial action are being undertaken, routine' control methods are planned and installed to make sure that the improvements secured are rnaintained. There are four main features of the approach that merit emphasis.

First, the audit deals with all those factors that affect the final quality of the product and is not confined to the normal laboratory tests, which give only part of the picture. The common difficulty in doing this is that there are so many factors that the data become unmanageable. The problem is alleviated in this method by building the data around investigation of one quality indicator. For spinning, doubling, and weaving, the number of 'end-down' is used.

It is naturall much easier to study interrelationships betweeen different factors if some sort of framework is available. The position is also eased by the use of very simple-methods of quality assessment, live the use of the +; ±; - scalc to evaluate certain characteristics. Obviously, other and more precise methods are employed, but generally the simple methods are sufficient to reveal processing or technical weaknesses.

The problem is the general one of trying to comply with the aims, which are sometimes mutually inconsistent, of keeping the data down to manageable proportions, getting them to be sufficiently accurate to be significant, and doing the job properly quickly with the kind of observer available. Since data are collected on all the factors affecting quality, the ensuing report thus provides a complete account of the quality position at one point in time. The interrelationship and interdependence of the various factors are clearly revealed, and such inconsistencies as may be present are highlighted.

Managements are therefore able to see the entire picture and can then use this to establish a quality policy. Secondly, the audit brings together the different groups of people who have. special responsibility for quality control. All learn to understand the contribution made by each other. In particular, the co-operation between laboratory and mill staff encouraged by the audit is likeiy to be continued afterwards. Many of the data-collection methods used can be adapted by the quality-control staff to help the supervisor to keep a check on the quality aspects of machine and operative performance.

The third feature is that of carrying out the whole survey with the mill's own staff. This helps both in the gathering of data and in the acceptance of results. Since all levels of mill staff have played some part in the final report, there is much greater likelihood that action to remedy particular faults win be taken.

Inevitably, the approach or method is aimed at helping the mill to help itself. Finally, the mill is being compared with itself. Although technological standards are used for comparison with regard to such characteristics as count, irregularity, etc., on many aspects none are available. In any case, to use such standards would be very unrealistic, especially when an older mill is being considered. Mill standards, however, are always computed.

For example, if one group of frames is significantly worse than the rest in some respect, then attention can be directed to those frames, even though there are no 'technological standards' available for frames of that age and condition.

Even if the long-term solution lies in re-equipment, some standards are essential to ensure that an the frames are brought to a common level, however low this might be.

## Control's Framework

In a publication as general as this, the intention is not to describe particular control procedures but rather to provide some indication of the anatomy of control. Furthermore, although the many factors are interrelated, for ease of presentation they have been considred under three headings: Materials Machines, and Operatives.

## Materials

In the beginning, it is essential to check whether the raw material, yarn, or cloth purchased conforms to the specification laid down. The significance of this aspect of control varies markedly from one section of the industry to another. For example, in cotton spinning it is feasible and desirable' to make on the raw cotton a range of tests covering such features as staple length, maturity, percentage short fibre, fibre strength and fineness, and trash content. The whole of the subsequent processing will be affected by the results of such tests. On the other hand, the weaver is often in the position of being unable to carry out more than the most rudimentary check on the standard of the incoming yarn, especially where beams are concerned.

Unless the small sample reveals a major fault, the usual objective is not so much to make between good or bad batches put, over a period of time, to select good suppliers. Even apart from the sample-size problem, there is the difficulty of the characteristics to be tested.

This is particularly so in weaving where the factors that determine the performance of a certain set in the loom are just as likely to be found in the way it was prepared as in the inherent qualities of the yarn. After the initial raw-material testing, the material may be tested after or during subsequent processing. These tests are to give both a check on the running of the process and a basis for action where results differ from the standard. These checks are an important part of control in such process industries as spinning, rather than in assembly ones such as weaving or knitting. For example, the regularity of a drawframe sliver may be tested on a daily basis. The over-all level provides an indication of the processing up to and including the drawframe process.

The existence of certain periodicities of varying wave lengths also provides an indication of the areas where action needs to be taken. In dealing with the final stage of material testing, i.e., the outgoing finished product, the aims of the procedures employed are three-fold. First they serve to give a check on the characteristics of the product to see whether they similar to customer standards. The check may result in the despatch of the cloth or yarn, etc., as satisfactory or it may lead to its being withheld. Between these two extremes, there are an infinite number of possible courses of action, which may comprise price allowances. Where the product is supplied to customers with varying acceptance standards, the final testing may be used to provide a basis for allotting the product to the different customers.

Secondly, the degree of conformance of the final product to the standard laid down may be regarded as an indication of the effectiveness of quality control in the mill; for instance, with such features as the variation in count, which reflects the eventual effectiveness of a series of controls at various stages of processing. Thirdly, the testing of the product can be regarded as a check on the final process.

The regularity of yarn, for instance, taken in conjunction with that of the roving fed to the ringframe, connotes not only an important feature of the final product but also the degree of control in the ringroom.

**Controlling Procedures**

Theoretically, were it feasible to setup clearly the exact causes of variations from standard by examining the product, there would be no need to consider control procedures other than those related to material and product testing. Practically, however, this is not feasible, and, even if it were, it does not necessarily follow that this would be the best thing to do. This is because the type of control procedure affects the likelihood that action will be taken on the results. For example, if yarn is examined and periodicities are revealed, it is feasible to conclude that an eccentric roller is causing the fault.

At the same time, there may be a multitude of causes that have contributed, and the most direct way is to examine the rollers with an eccentricity gauge. This has the added benefit that it can be performed by the overlooker himself. Regarding machines, it can be argued that it is first essential to check whether the postulated quality characteristics are consistent with the capabilities of the machine.

This corresponds to the problem of selecting materials that can be expected to give certain characteristics in the final product. For example, just as it is impossible to spin yarns of a certain count and strength from some cottons, it is equally impossible to produce yarn of a certain level of regularity from machines of certain age and design.

There is also the question of whether it is feasible to produce a some quality by processing the material through a limited sequence of machines. These last two factors are interrelated with the process capabilities of the individual machines making up that sequence.

The action associated with these checks and process-capability studies will depend on both the results and the time factor. Small deviations from standard may lead to adjustment in the succeeding process or to changing the feeding procedures at the machine tested.

Machine checks initially comprise ensuring that the machine has been correctly set up, i.e., that the speeds and settings are consistent with obtaining the desired characteristics in the product. It is also essential to check periodically that these speeds and settings remain as laid down. Checks should be made to ensure that the condition of the machines is being maintained at a level consistent with quality needs. This provides a good example of how various levels of management are affected by the results of a particular control procedure. It may divulge the need for relatively minor action, e. g., the fitting of new traveller cleaners in ring-spinning frames. The decision here can be made at departmental level. It may reveal the necessity for changes in the stipulated maintenance programme, e.g., increasing scouring frequencies. The proper action here involves. the mill manager in that the effects of such a change will affect the balance of production in all departments.

## QUALITY CONTROL OPERATIVES

Regarding operatives, there are really three chief areas of quality-control checks or procedures. Each must be covered to make sure that quality reaches, and is maintained at, the required level. First, it is necessary to establish whether the operatives possess the level of ability needed.

Associated closely with this aspect of control, which is more accurately part of selection, is the question of training the operatives once they have been opted, It is very easy to do this with new operatives if systematic training procedures are adopted, but much more difficult with experienced Operatives whose bad quality habits are deeply rooted.

Quality practices are by no means standardized, and it follows that each operative should be clearly informed of what he or she is supposed to do in very considerable detail. This can be discussed by the possible ways of carrying out the apparently straightforward task of creeling in a can at the drawframe. There are two normal ways of doing this:

- Break the sliver and piece it to the beginning of a new can, a remnant being left in the bottom of the old one.
- Piece start or the full can to the end of the preceding can; this comprises removing the remainder of the sliver from the old can while the frame is running, turning it over, and piecing the end to the beginning of the full can; and

The piece that is created as an out come of the adoption of the second method can, in turn, be disposed of in one of several ways:

- It can be placed in a special 'piece can'; or
- It can be pieced to the top of the full can later;
- The can comprising it can be transferred to the front of the preceding head and used instead of an empty can at doff.
- It can be discarded as, vaste;

The quality of the product is explicitly affected by the creation and use of these pieces. If the first or third means of disposal is employed, two further piecings are necessary to put the piece back into the system.

The 'piece-can' method has an additional and obvious detrimental effect on quality. The last method needs only one join to put a piece back into the system, but, if the practice is continued, the piece,will remain in the can for a long period of time. This particular aspect has been given in some detail to illustrate the fact that there are often several ways of performing a relatively straight forward task in an individual situation.

To make sure that the correct one is used involves not only a detailed breakdown of the task but clear instructions on each part of it, followed by regular checks to ensure that the stipulated procedure is adhered to. The second area requiring attention is the work allocation. Does it allow the achievement of the quality standards? This must be checked, for it cannot be assumed to be correct.

In this regard, the role of techniques such as work study is important. For instance, it is just as significant to calculate the amount of cleaning necessary as it is to discover the amount of time it takes to clean. Too often cleaning frequencies bear no relation to the standard of cleanliness of machines required for a given level of product cleanliness. The work-study officer is not liable for the decision on such frequencies: he merely takes them into account in calculating the over-all work-load.

The precision in timing a special operation is likely, however, to give an aura of objectivity, which conceals the fact that decisions on the nature and extent of cleaning are frequently made quite unsystematically. If one considers work organization in the broader aspect, there seems little doubt that some systems of work allocation do adversely affect the quality of the product. For instance, it is customary in ring spinning for the operative to be for patrolling the spinning frames both to repair broken ends and to replenish the empty supply packages.

Some work-loading systems have aimed at separating these two jobs. Theoretically, there should be no overlap between the duties allocated to each group. In practice it is by no means so simple. The division of responsibilities tends to have an adverse effect on quality, with respect both to the individual operative, who finds other people are playing a part in affecting the processing conditions, and to the supervisor, who finds it difficult to allocate responsibility. This is not necessarily to imply that these disadvantages may not be more than offset by increased machine efficiency, etc.

In general terms then, it is wrong to consider standards of operative performance without regard to the particular work organization in which they are expected to operate. This should be interprcted in the widest sense and should not be confined to the division of duties.

## PROTEIN FIBRES

The protein fibres are formed by natural animal sources through condensation of a-amino acids to form repeating polyamide units with various substituents on the a-carbon atom. The sequence and type of amino acids making up the individual protein chains contribute to the overall properties of the resultant fibre. Two major classes of natural protein fibres exist and include keratin and secreted fibres. In general, the keratin fibres are proteins highly crosslinked by disulfide bonds from cystine residues in the protein chain, whereas secreted fibres tend to have no crosslinks and a more limited array of less complex amino acids.

The keratin fibres tend to have helical portions periodically within protein sequence, whereas the secreted fibre protein chains are arranged in a linear pleated sheet structure with hydrogen bonding between amide groups on adjacent protein chains. Keratin fibres are extremely complex in structure and include a cortical cell matrix surrounded by a cuticle sheath laid on the surface as overlapping scales.

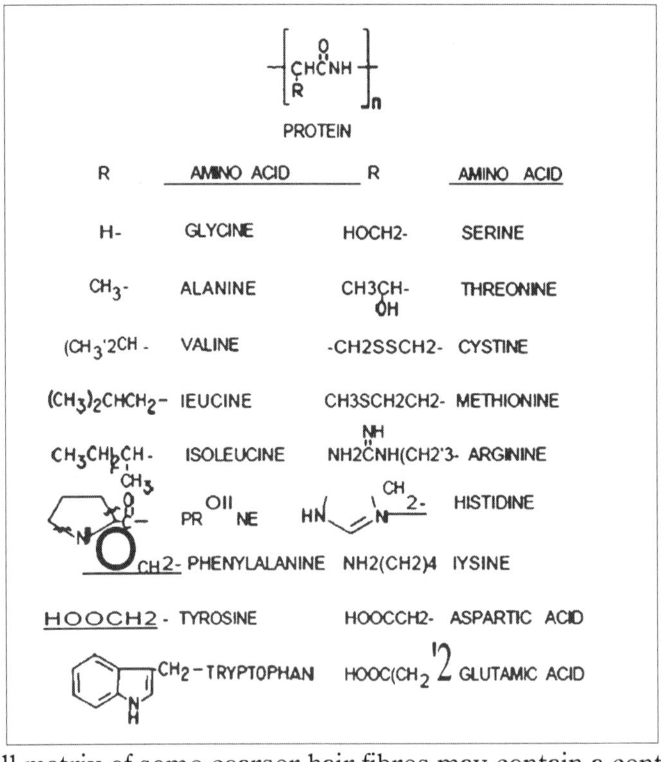

$$\left[ -\underset{R}{C}H\overset{O}{\overset{\|}{C}}NH- \right]_n$$

PROTEIN

| R | AMINO ACID | R | AMINO ACID |
|---|---|---|---|
| H- | GLYCINE | HOCH2- | SERINE |
| $CH_3$- | ALANINE | CH3CH(OH)- | THREONINE |
| $(CH_3)_2CH$- | VALINE | -CH2SSCH2- | CYSTINE |
| $(CH_3)_2CHCH_2$- | IEUCINE | CH3SCH2CH2- | METHIONINE |
| $CH_3CH_2CH(CH_3)$- | ISOLEUCINE | NH2C(=NH)NH(CH2)3- | ARGININE |
| [ring structure] | PROLINE | [ring structure] CH2- | HISTIDINE |
| [ring structure] CH2- | PHENYLALANINE | NH2(CH2)4 | IYSINE |
| HOOCH2- | TYROSINE | HOOCCH2- | ASPARTIC ACID |
| [ring structure] CH2- | TRYPTOPHAN | $HOOC(CH_2)_2$ | GLUTAMIC ACID |

The cell matrix of some coarser hair fibres may contain a center cavity or medulla. The keratin fibres are round in cross section with an irregular crimp along the longitudinal fibre axis which results in a bulky texturised fibre. Secreted fibres are much less complex in morphology and often have irregular cross sections. Other fibrous protein materials include azlon fibres, spun from dissolved proteins, and a graft protein-acrylonitrile matrix fibre. In general, protein fibres are fibres of moderate strength, resiliency, and elasticity. They have excellent moisture absorbency and transport characteristics. They do not build up static charge. While they have fair acid resistance, they are readily attacked by bases and oxidising agents. They tend to yellow in sun light due to oxidative attack. They are highly comfortable fibres under most environmental conditions and possess excellent aesthetic qualities.

## WOOL

Wool is a natural highly crimped protein hair fibre derived from sheep. The fineness and the structure and properties of the wool will depend on the variety of sheep from which it was derived. Major varieties of wool come from Merino, Lincoln, Leicester, Sussex, Cheviot, and other breeds of sheep. Worsted wool fabrics are made from highly twisted yarns of long and finer wool fibres, whereas woolen fabrics are made from less twisted yarns of coarser wool fibres.

### Structural Properties

Wool fibres are extremely complex, highly crosslinked keratin proteins made up of over 17 different amino acids. The amino acid content and sequence in wool varies with variety of wool. The average amino acid contents for the major varieties of wool are given in table.

**Table. Amino Acid Contents in Wool Keratins.**

| Amino Acid | Content in Keratin (g/100 g Wool) |
|---|---|
| Glycine | 5-7 |
| Alanine | 3-5 |
| Valine | 5-6 |
| Leucine | 7-9 |
| Isoleucine | 3-5 |
| Proline | 5-9 |
| Phenylalanine | 3-5 |
| Tyrosine | 4-7 |
| Tryptophan | 1-3 |
| Serine | 7-10 |
| Threonine | 6-7 |
| Cystine | 10-15 |
| Methionine | 0-1 |
| Arginine | 8-11 |
| Histidine | 2-4 |
| Lysine | 0-2 |
| Aspartic acid | 6-8 |
| Glutamic acid | 12-17 |

The wool protein chains are joined periodically through the disulfide crosslinked cystine, a diamino acid that is contained within two adjacent chains. About 40 per cent of the protein chains spiral upon themselves and internally hydrogen bond to form an a-helix. Near the periodic cystine crosslink s or at points where proline and other amino acids with bulky groups occur along the chain, the close packing of chains is not possible and a less regular non-helical structure is observed.

The crosslinked protein structure packs and associates to form fibrils, which in turn make up the spindle shaped cortical cells which constitute the cortex or interior of the fibre. The cortex is made up of highly and less crosslinked ortho and para cortex positions. The cortex is surrounded by an outer sheath of scalelike layers or cuticle, which accounts for the scaled appearance running along the surface of the fibre.

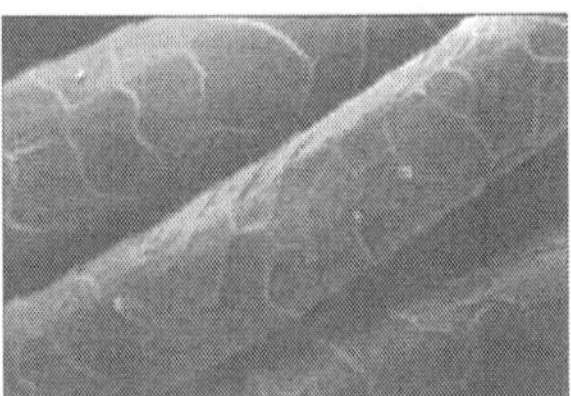

**Fig.** Wool.

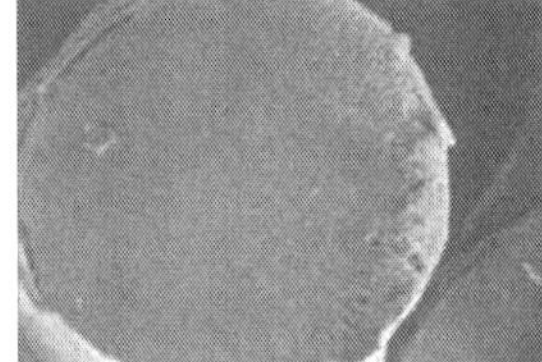

**Fig.** Wool Cross-section.

**Physical Properties**

Wool fibre s possess low to moderate strength with tena cities of 1-2 g/ d dry and 0.8-1. 8 g/ d wet. Elongations at break vary from 25 per cent to 40 per centdry and 25 per cent to 60 per centwet. At 2 per cent extension, wool shows 99 per centrecovery, and even at 20 per centextension a recovery as high as 65 per centis observed. Wool fibres have excellent resiliency and recover readily from deformation except under high humidities. The stiffness of wool varies according to the source and the diameter of the individual fibres.

The moisture regain of wool is very high and varies between 13 per cent and 18 per cent under standard conditions. At 100 per cent RH, the regain approaches 40 per cent. Wool fibres have specific gravities of 1.28-1.32. Wool is insoluble in all solvents except those capable of breaking the disulfide crosslinks, but it does tend to swell in polar solvents. Wool is little affected by heat up to 150°C and is a good heat insulator due to its low heat conductivity and bulkiness, which permits air entrapment in wool textile structures. At moderate humidities, wool does not build up significant static charge.

**Chemical Properties**

Wool is resistant to attack by acids but is extremely vulnerable to attack by weak bases even at low dilutions. Wool is irreversibly damaged and coloured by dilute oxidising bleaches such as hypochlorite. Reducing agents cause reductive scission of disulfide bonds within the wool, eventually causing the wool to dissolve. Under controlled conditions, reducing agents can be used to partially reduce the wool and flat set or set permanent pleats in the wool. Unless chemically treated, wool is susceptible to attack by several species of moths able to dissolve and digest wool fibres. Wool is quite resistant to attack by other biological agents such as mildew. Wool is attacked by short wavelength ultraviolet light, causing slow degradation and yellowing. On heating, wool degrades and yellows above 150 °C and chars at 300 °C.

**End-Use Properties**

Wool varieties include Merino, lincoln, leicester, Sussex, Cheviot, Ramboullett, and Shetland, as well as many others. Wool is a fibre of high to moderate luster. Fabrics of wool posses s a soft to moderate hand and exhibit good drapability. Wool fibres are highly absorbent and have excellent moisture transmission properties. The low to moderate strength of wool fibres is compensated for by its good stretch and recovery properties. Wool is fairly abrasion resistant and does not tend to form pills due to its low strength.

It resists wrinkling except under warm, moist conditions. Its crease retention is poor unless creases have been set using chemical reducing agents. Wool is attacked by alkalies and chlorine bleaches and is progressively yellowed by the short ultraviolet wavelengths in sunlight. Wool dyes readily, and the dyed wools exhibit good colourfastness. Owing to its felting action, wool cannot be laundered in hot water with agitation, but it can be dry-cleaned or washed in warm water with a mild detergent if no agitation is used. Due to its affinity for water, wool is slow drying. Wool may be ironed at 150 °C or below without steaming. Wool is a self-extinguishing fibre and burns very slowly even in contact with a flame. It has an lOI of 25. Wool is extensively used in textile applications where comfort and aesthetics are important. It is used in men's and women's apparel, outer wear and cold weather clothing, suits, blankets, felts, and carpeting. It is often used in blends with cellulosic and man-made fibres.

**SILK**

Silk is a natural protein fibre excreted by the moth larva Bombyx mori, better known as the common silkworm. Silk is a fine continuous monofilament fibre of high luster and strength and is highly valued as a prestige fibre. Because of its high cost, it finds very 1imited use in textiles. A minor amount of wild tussah silk is produced for specialty items.

Attempts have been made to commercialise silk from spiders over the years, but all ventures have met with failure. Domestic and wild silks are essentially uncrosslinked and relatively simple in amino acid composition compared to the keratin fibres.

The properties for silk listed here are for silk formed by Bombyx mori moth larvae.

Liquid silk protein is extruded from two glands in the head of the silkworm. The fibres emerge from a common exit tube or spinneret and harden into a single strand by a protein gum called sericin.

The completed silk cocoons are soaked in hot water to loosen the sericin, and the silk filaments are unwound. After unwinding, the silk filaments are washed in warm detergent solutions to remove the sericin. The fibroin silk fibres are more simple in structure than keratin and are composed predominantly of glycine, alanine, tyrosine, and serine. The average range of composition for silk is given in table.

With no cystine present in the fibroin protein, little crosslinking is observed between protein chains. The degree of polymerisation of silk fibroin is uncertain, with DPs of 300 to 3000+ having been measured in different solvents. In the absence of crosslinks and with limited bulky side chains present in the amino acids, fibroin molecules align themselves parallel to each other and hydrogen bond to form a highly crystal line and oriented "pleated-sheet" or "beta" structure.

Silk fibres are smooth surfaced and translucent with some irregularity in diameter along the fibre. The fibres are basically triangular in cross section with rounded corners.

**Table. Amino Acid Contents in Fibroin.**

| Amino Acid | Content in Fibroin (g/IOO g Fibroin) |
|---|---|
| Glycine | 36-43 |
| Alanine | 29-35 |
| Tyrosine | 10-13 |
| Serine | 13-17 |
| Valine | 2-4 |
| Leucine | 0-1 |
| Proline | 0-1 |
| Phenylalanine | 1-2 |
| Tryptophan | 0-1 |
| Threonine | 1-2 |
| Cystine | 0 |
| Methionine | 0 |
| Arginine | 0-2 |
| Histidine | 0-1 |
| Lysine | 0-1 |
| Aspartic acid | 1-3 |
| Glutamic acid | 1-2 |

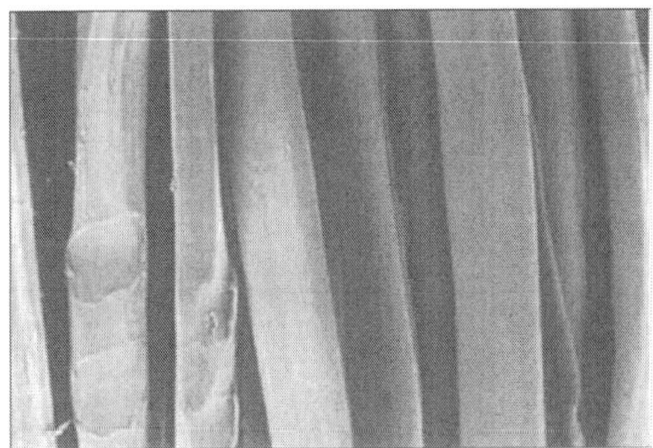

**Fig.** Silk.

### Physical Properties

Silk fibres are strong with moderate degrees of recovery from deformation. Silk has a dry tenacity of 3-6 g/d and a wet tenacity of 2.5-5 g/d. Silk exhibits a recovery of 90 per cent from 2 per cent elongation and of 30 per cent-35 per cent from 20 per cent elongation. Silk fibres are moderately stiff and exhibit good to excellent resiliency and recovery from deformation, depending on temperature and humidity conditions. Silk has a specific gravity of 1.25-1.30 and a moisture regain of 11 per cent under standard conditions.

Silk is soluble in hydrogen bond breaking solvents such as aqueous lithium bromide, phosphoric acid, and cuprammonium solutions. It exhibits good heat insulating properties and is little affected by heat up to 150°C. Silk has moderate electrical resistivity and tends to build up static charges.

### Chemical Properties

Silk is slowly attacked by acids but is damaged readily by basic solutions. Strong oxidising agents such as hypochlorite rapidly discolour and dissolve silk, whereas reducing agents have little effect except under extreme conditions. Silk is resistant to attack by biological agents but yellows and loses strength rapidly in sunlight. Silk is often weighted with tin and other metal salts. These salts make silk even more sensitive to light-induced oxidative attack. Silk undergoes charring and oxidative decomposition when heated above 175°C in air over a prolonged period of time.

### End-Use Properties

Silk possesses a combined set of aesthetic properties that make it useful for high-fashion luxury textile goods. Silk has a high luster and is translucent. Silk fabrics have pleasing appearance and drapability, and a characteristically pleasing crisp hand. Silk is highly moisture absorbent and has well to excellent resistance to wrinkling. It is a moderately strong fibre with moderate recovery properties. It exhibits fair abrasion resistance and good resistance to pilling.

Silk is sensitive to chlorine bleaches and to alkalies and is easily damaged by sunlight. The fibre may be dyed with a wide variety of dyes to give dyed fibres with high colourfastness. Silk may be laundered under mild, non-alkaline conditions and dry cleans readily. Because of its high affinity for water, it dries slowly but may be dried or ironed safely up to 150°C. Silk burns slowly and self-extinguishes when removed from a flame. Silk is used extensively in luxury fabrics and apparel and home furnishings, and in accessories such as scarfs.

## OTHER NATURAL AND REGENERATED PROTEIN FIBRES

*The other major hair fibres include:*

- Mohair,
- Cashmere,
- Llama,
- Alpaca, and
- Vicuna, as well as many others.

Regenerated Azlon fibres are derived from soluble proteins that can be spun into fibres, insolubilised, and regenerated. Soluble proteins also may be grafted to form a copolymer, dissolved, and then spun into fibres.

### Mohair

Mohair is a very resilient hair fibre obtained from the angora goat. The two primary classifications for mohair are the finer kid mohair and the coarser adult mohair. In many respects, mohair resembles wool in structure and possesses properties including the characteristic scale structure of the fibre. The average length of mohair fibres is longer than wool, with 4-12 inch fibre lengths being typical.

It is a much stronger fibre than wool, but its other tensile properties resemble wool. Mohair is remarkably resistant to wear and is used in applications where such durability is essential. Mohair possesses a beautiful natural luster and is used in blends to provide luster. Mohair fibres also provide a characteristic, resilient, and slightly scratchy hand in even small quantities blended with other fibres.

### Cashmere

Cashmere is the fine, soft inner coat of down obtained from the cashmere goat found on the inner plateaus of Asia. In many ways the properties of cashmere resemble those of wool, but cashmere fibres are extremely fine and soft compared to wool. Cashmere is used in luxury applications where a soft, warm, fine fibre with beautiful drape is desired.

### Llama, Alpaca, and Vicuna

These fibres come from a group of related animals found in South America. They are fine fibres that are white to tan and brown in colour. They are longer than most wool fibres and generally stronger, with a finer scale structure. They are generally used only in the expensive luxury items of textiles and apparel.

### Regenerated Protein Fibres

Azlon is the generic name given to manufactured fibres composed of a regenerated natural protein. Azlon is produced by dissolving proteins like casein from milk, soya bean protein, and zein from corn in dilute alkali and forcing these solutions through a spinneret into an acid-formaldehyde coagulating bath. Many of the properties of these fibres resemble the natural protein fibres, but they suffer from low dry and wet strength and sensitivity to alkalies. Although no longer produced in the U.S., azlon fibres are produced in Europe and used in blends with natural and man-made fibres.

### Protein-Polyacrylonitrile Graft Copolymer

A fibre consisting of a copolymer of casein protein grafted with 40 per cent-75 per cent acrylic monomers, of which at least half is acrylonitrile, has been developed in Japan under the tradename Chinon. The casein dissolved in aqueous zinc chloride and grafted with acrylonitrile is wet or dry spun into fibres. The fibre has a tenacity of 3.5-5 g/d dry and 3-4.5 g/d wet and an elongation at break of 15 per cent-25 per cent wet or dry. It recovers 70 per cent from 5 per cent elongation. The fibre has a moisture regain of 4.5 per cent-5.5 per cent and a specific gravity of 1.22. It dyes readily with acid dyes, but basic and reactive dyes can be used also. The fibre is marketed as a substitute for silk.

## NYLON AND POLYAMIDE FIBRES

The polyamide fibres include the nylons and the aramid fibres. Both fibre types are formed from polymers of long-chain polyamides. In nylon fibres less than 85 per cent of the polyamide units are attached directly to two aromatic rings, whereas in aramid fibres more than 85 per centof the amide groups are directly attached to aromatic rings.

The nylons generally are tough, strong, durable fibres useful in a wide range of textile applications. The fully aromatic aramid fibres have high temperature resistance, exceptionally high strength, and dimensional stability. The number of carbon atoms in each monomer or comonomer unit is commonly designated for the nylons. Therefore the nylon with six carbon atoms in the repeating un it would be nylon 6 and the nylon with six carbons in each of the monomer units would be nylon 6,6.

### NYLON 6 AND 6,6

Nylon 6 and 6,6 are very similar in properties and structure and therefore will be described together. The major s tructural difference is due to the placement of the amide groups in a continuous head-to-head arrangement in nylon 6, whereas in nylon 6,6 the amide groups reverse direction each time in a head-to-tail arrangement due to the differences in the monomers and polymerisation techniques used:

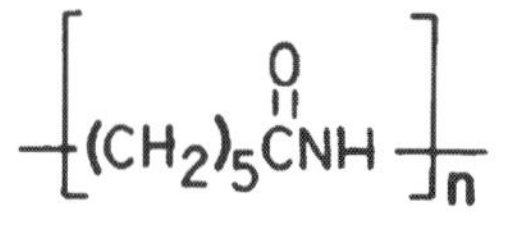

$$\left[ (CH_2)_5\overset{O}{\overset{\|}{C}}NH \right]_n$$

NYLON 6

$$\left[ \overset{O}{\overset{\|}{C}}(CH_2)_4\overset{O}{\overset{\|}{C}}NH(CH_2)_6NH \right]_n$$

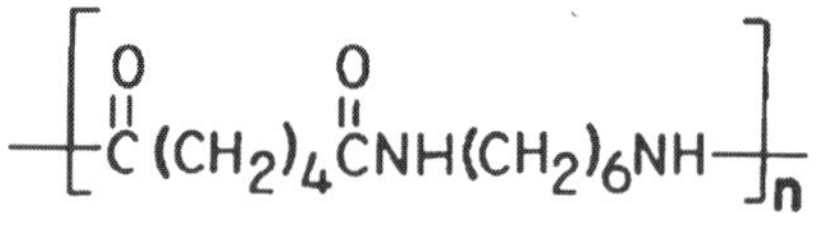

NYLON 6.6

Nylon 6,6 was developed in the United States, whereas nylon 6 was developed in Europe and more recently Japan. These nylon polymers form strong, tough, and durable fibres useful in a wide variety of textile applications. The major differences in the fibres are that nylon 6,6 dyes lighter, has a higher melting point, and a slightly harsher hand than nylon 6.

**Structural Properties**

Nylon 6 is produced by ring-opening chain growth polymerisation of caprolactam in the presence of water vapor and an acid catalyst at the melt. After removal of water and acid, the nylon 6 is melt spun at 250°C260° C into fibres. Nylon 6,6 is prepared by step growth polymerisation of hexamethylene diamine and adipic acid. After drying, the nylon 6,6 is melt spun at 280°C-290°C into fibres. Both nylon 6 and 6,6 are drawn to mechanically orient the fibres following spinning.

The degree of polymerisation of nylon 6 and 6,6 molecules varies from 100 to 250 units. The polyamide molecular chains lay parallel to one another in a "pleated sheet" structure similar to silk protein with strong hydrogen bonding between amide linkages on adjacent molecular chains. The degree of crystallinity of the nylon will depend on the degree of orientation given to the fibre during drawing.

Nylon fibres are usually rodlike with a smooth surface or trilobal in cross section. Multilobal cross sections and other complex cross sections are also found. Side-by-side bicomponent fibres with a round cross section are also formed. Following orientation and heating, these bicomponent fibres form small helical crimps along the fibre to give a texturised fibre useful in many applications including women's hosiery. Cantrece nylon is such a fibre. Also, sheath-core nylons with differing melting characteristics are formed for use as self-binding fibres for non-wovens.

**Fig.** Nylon 6,6, Round.

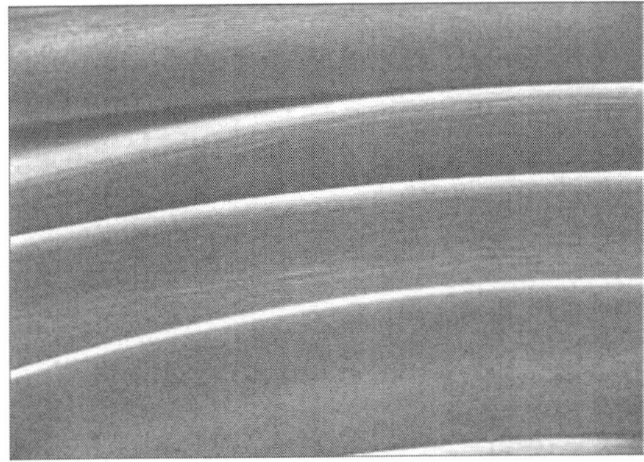

**Fig.** Nylon 6,6, Trilobal.

## Physical Properties

Nylon 6 and 6,6 fibres are strong, with a dry tenacity of 4-9 g/d and a wet tenacity of 2.5-8 g/d. These nylons have elongations at break of 15 per cent-50 per cent dry, which increase somewhat on wetting. Recovery from stretch deformation is very good, with 99 per cent recovery from elongations up to 10 per cent.

The nylons are stiff fibres with excellent resiliency and recovery from bending deformation. They are of low density, with a specific gravity of 1.14. They are moderately hydrophilic with a moisture regain of 4 per cent-5 per cent under standard conditions and a regain of 9 per cent at 100 per cent relative humidity. Nylon 6 and 6,6 are soluble in hydrogen bond breaking solvents such as phenols, 90 per cent formic acid, and benzyl alcohol. They have moderate heat conductivity properties and a re unaffected by heating below 150°C. The nylons have a high resistivity and readily build up static charge.

## Chemical Properties

The nylons are fairly resistant to chemical attack. They are attacked by acids, bases, and reducing and oxidising agents only under extreme conditions. They are unaffected by biological agents, but at elevated temperatures or in the presence of sunlight they will undergo oxidative degradation with yellowing and loss of strength.

## End-Use Properties

Nylon 6 and 6,6 are marketed under several trade names. In fact, nylon was originally used as a trade name for duPont polyamide fibre, but as a result of common usage the term nylon came to be the generic term for these polyamide fibres.

*Common trade names for these nylons include:*

- Anso,
- Antron,
- Cadon,
- Cantrece,
- Cordura,
- Caprolan,
- DuPont nylon, and
- Enkalure.

Nylon 6 and 6,6 are extremely strong fibres with excellent recovery and resiliency properties. Nylon fabrics have a fair hand, with nylon 6 having a somewhat softer hand than nylon 6,6. Nylon fabrics have high luster unless delustered. They have moderate to excellent draping properties depending on the denier of the fibre. Unless bleached and dyed, nylons have a slightly yellow colour. The nylons are wrinkle resistant and have good crease resistance if heat set. The fibre is tough and has good abrasion resistance. In fact, nylon is reported to abrade other fibres in a fibre blend. The fibre has good resistance to household chemicals but exhibits poor resistance to attack by sunlight unless treated with antioxidants. The fibres have excellent dyeability with excellent colourfastness properties.

Nylon 6 is somewhat deeper dyeing than nylon 6,6. The fibre has good laundering and dry cleaning properties but tends to scavenge dyes bleeding from other fibres.

The fibres have moderate moisture uptakes and dry readily at temperatures up to 150°C. Nylon 6 can be safely ironed up to 140°C, whereas nylon 6,6 can be safely ironed up to 180°C. The nylons are less flammable than cellulosics, with an LOI of 20. They melt and drip and tend to self-extinguish on burning. Nylons are extensively used in hosiery, lingerie, underwear, sweaters, and other knitted goods. They have been used extensively in light and sheer apparel articles such as windbreakers.

*Nylons are used in home furnishings and industrial applications including:*

- Carpets,
- Upholstery,
- Tie cord,
- Parachutes,
- Sails,
- Ropes,
- Thread, and
- Outdoor wear.

## ARAMID FIBRES

The aramid polyamide fibres are formed from a long chain of synthetic polyamides in which at least 85 per cent of the amide linkages are attached to aromatic rings. These essentially fully aromatic polyamides are characteristically high melting and have excellent property retention at high temperatures and excellent durability. They are unaffected by moisture and most chemicals and are inherently flame retardant. The fibres have high strength and can be used in a number of unique high-strength applications.

**Structural Properties**

Aramids are formed through step growth polymerisation of aromatic diacid chlorides with aromatic diamines in a polar aprotic solvent such as N,N-dimethylformamide (DMF) to a DP of 100-250. The meta- and para-substituted benzene dicarboxylic acid chlorides and diamines are characteristically used for aramid fibres presently in production, but other fully aromatic ring systems are possible future sources of aramid polymers for fibres:

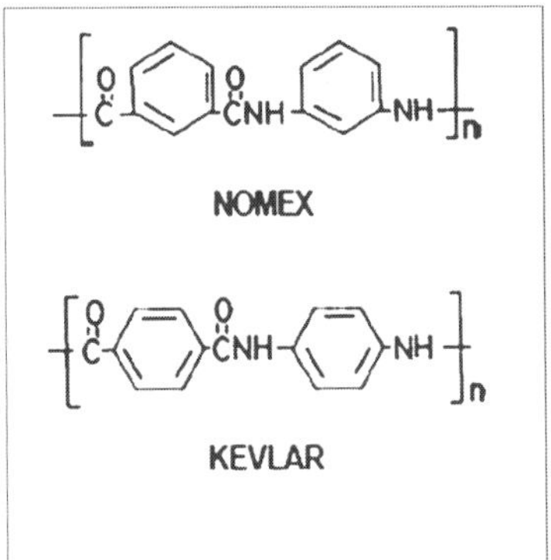

NOMEX

KEVLAR

ARAMID

The resultant aramid polymers are spun in suspension through a spinneret into hot air or a coagulating of both to form the fibres, followed by fibre stretching and orientation. The aromatic units along aramid polyamide chains confer stiffness to the polymeric chains and limit their flexibility and mobility. Hydrogen bonding between amide groups on adjacent chains and extremely strong van der Waals interactions between aromatic rings planar to adjacent aromatic rings provide a tightly- packed, strongly held molecular structure and account for the strength and thermal stability of the aramids. The aramids are usually spun in round or dumbbell cross section.

### Physical Properties

The aramids are the strongest of the man-made fibres, with the strength depending on the polymer structure, spinning method used, and the degree of orientation of the fibre. Dry tenacities of 4-22 g/d are found for the aramids, and wet tenacities of 3-18 g/d have been recorded. The elongation at break of the fibres decreases with increasing tenacity over a range of 3 per cent-30 per centdry and wet.

Recovery from low elongations of 5 per cent or less is 98 per cent-100 per cent. The fibres are stiff, with excellent resiliency and recovery from bending deformation. The specific gravities of these fibres vary from 1.38 to 1.44. The aramids have a moisture regain comparable to the other polyamide fibres and are in the 3.5 per cent-7 per cent range. The fibres are swollen and dissolved in polar aprotic solvents or strong acids. The aramids have high heat and electrical resistivities, have excellent insulative capabilities, and are unaffected by heat up to 250°C.

### Chemical Properties

The aramids are extremely resistant to chemical or biological attack. Only under extreme conditions and at elevated temperatures will concent rated acids or strong oxidising agents attack aramids. Aramid fibres undergo initial oxidative attack in sunlight, causing discolouration and slight strength loss, but further exposure has little additional effect. The aramids also act as effective screens to high-energy nuclear radiation due to their ability to trap and stabilise radical and ionic species induced by radiation. The aramids undergo oxidative degradation on prolonged heating above 370°C.

### End-Use Properties

Common trade names for aramid fibres include Nomex and Kevlar. Aramid fibres are extremely strong and heat resistant. Fabrics from the aramids have a high luster with a fair hand and adequate draping properties. The fibres are light yellow unless bleached and exhibit moderate moisture absorption characteristics. The fibres recover readily from stretching and bending deformation and are extremely abrasion resistant. They do tend to pill due to the high strength of the fibre. The aramids are extremely resistant to attack by household chemicals and exhibit good resistance to sunlight. The fibres are difficult to dye except by special dyeing techniques with disperse dyes. Dyed aramids are reasonably colourfast.

They have good launderability and dry-cleanability and are moderately easy to dry. The fibres are extremely heat stable and can be ironed up to 300°C. The fibres are of low flammability and self-extinguish on removal from a flame. They have an LOI of 30. Aramid fibres are used in fabric applications where high strength and low flammability are important, including protective clothing, electrical insulation, filtration fabrics, specialised military applications, and tire cord.

## OTHER POLYAMIDES

Several other polyamides have been introduced for use as fibres in specialty applications where certain combinations of properties are desired. The major specialty nylons include Qiana, nylon 4, nylon 11, nylon 6,10, and biconstituent nylon-polyester.

Qiana is a trade name for the luxury nylon fibre formed through step growth polymerisation of trans, transdimethane and a dibasic acid having 8-12 carbon atoms:

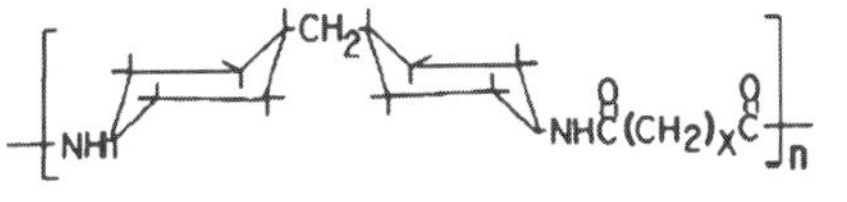

QIANA

Qiana resembles nylon 6 and 6,6 in many of its properties but also has a unique silklike texture, a lustrous appearance, excellent form retention, and an overall superior performance level which makes it useful in prestige textile products and in the sophisticated fashion apparel market. Qiana has a tenacity of 3-3.5 g/d and an elongation at break of 20 per cent-30 per cent. It is reported to recover from strain better than nylon 6,6 or polyester. It has a specific gravity of 1.03 and a moisture regain of 2 per cent-2.5 per cent, which are much lower values than nylon 6 or 6,6. Qiana melts at 275°C and is stable up to 185°C.

The cross section of the fibre is trilobal. It shows good chemical stability and is readily dyeable to fast colours. The excellent aesthetic qualities and drapability of Qiana have contributed greatly to its success.

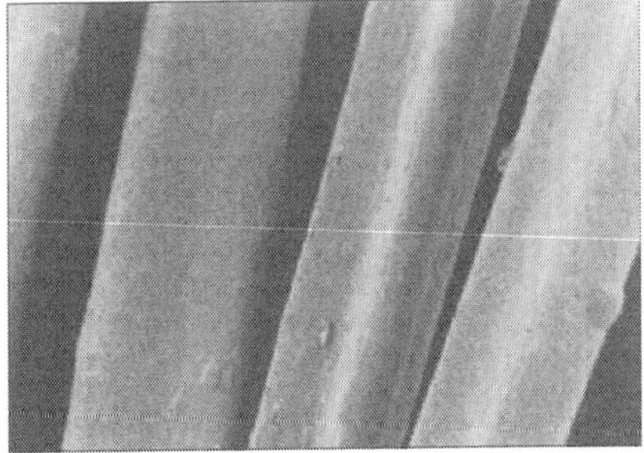

**Fig.** 6-3. Qiana

### Nylon 4

Nylon 4 is produced by polymerisation of pyrrolidone using carbon dioxide catalyst. Nylon 4 is melt spun to give a fibre of about 60 per cent crystallinity. The fibre is of moderate dry strength and slightly higher wet strength, with a higher specific gravity than nylon 6 and 6,6. The fibre has excellent elongation and recovery properties and a water regain comparable to cotton. Nylon 4 is easily laundered and can be dyed easily to give colourfast shades but has limited wrinkle recovery properties and is sensitive to hypochlorite bleaches. Although a major attempt was made to commercialise nylon 4, market conditions limited the penetration of this new fibre type and production was discontinued. Nylon 4 is no longer being produced.

### Nylon 11

Nylon II is produced by self-condensation of ll-aminoundecanoic acid in Europe and marketed under the name Rilsanite. It is melt spun into fibre and possesses essentially the same properties as nylon 6 and 6,6. It has high strength, low specific gravity, and low water regain. Nylon II has excellent electrical properties and is used in electrical products as well as in brush bristles, tire cord, lingerie, and hose.

### Nylon 6,10

Nylon 6,10 is produced by condensation polymerisation of he xame thylene diamine and sebacic acid and is melt spun into fibres. It resembles nylon 6 and 6,6 in many ways but has a lower moisture regain. It is primarily used in brush bristles.

### Biconstituent Nylon-Polyester

Biconstituent fibre of nylon 6 with polyester microfibrils dispersed throughout the fibre matrix has been marketed under the trade name Source. The fibre is reported to have unique optical and dyeing properties and a higher strength and lower regain than nylon 6,6 and is used primarily in carpets. A sheath-core bicomponent fibre containing a nylon 6 sheath and a polyester core has been reported al so. It is said to have properties that are intermediate between both fibres.

## POLYETHYLENE TEREPHTHALATE

Polyethylene terephthalate polyester is the leading man-made fibre in production volume and owes its popularity to its versatility alone or as a blended fibre in textile structures. When the term "polyester" is used, it refers to this generic type. It is used extensively in woven and knitted apparel, home furnishings, and industrial applications. Modification of the molecular structure of the fibre through texturising and or chemical finishing extends its usefulness in various applications. Polyester is expected to surpass cotton as the major commodity fibre in the future.

## POLYESTER FIBRES

Polyesters are those fibres containing at least 85 per cent of a polymeric ester of a substituted aromatic carboxylic acid including but not restricted to terephthalic acid and f-hydroxybenzoic acid. The major polyester in commerce is polyethylene terephthalate, an ester formed by step growth polymerisation of terephthalic acid and the diol ethylene glycol. Polyl, 4-cyclohexylenedimethylene terephthalate is the polyester of more limited usage and is formed through the step growth polymerisation of terephthalic acid with the more complex diol 1,4-cyclohexylenedimethanol. A third polyester fibre is actually the polyester ether PolY and is formed through step growth polymerisation of £-hydroxybenzoic acid with ethylene glycol; it is no longer in commercial production, however. The polyester fibres all have similar properties, are highly resilient and resistant to wrinkling, possess high durabilily and dimensional stability, and are resistant to chemical and environmental attack.

### Structural Properties

Polyethylene terephthalate is formed through step growth polymerisation of terephthalic acid or dimethyl terephthalate with ethylene glycol at 250°-300°C in the presence of a catalyst to a DP of 100-250:

$$\left[ -\overset{O}{\overset{\|}{C}}-C_6H_4-\overset{O}{\overset{\|}{C}}OCH_2CH_2O- \right]_n$$

POLYETHYLENE TEREPHTHALATE
POLYESTER

The resultant polymer is isolated by cooling and solidification and dried. Polyester fibres are melt spun from the copolymer at 250°-300°C, followed by fibre orientation and stretching.

**Fig.** Polyester.

The polyester molecular chains are fairly stiff and rigid due to the presence of periodic phenylene groups along the chain. The polyester molecules within the fibre tend to pack tightly and are held together by van der Waals forces. The polyesters are highly crystalline unless comonomers are introduced to disrupt the regularity of the molecular chains. Polyester fibres are usually smooth and rodlike with round or trilobal cross sections.

**Physical Properties**

Polyester from polyethylene terephthalate is an extremely strong fibre with a tenacity of 3-9 g/d. The elongation at break of the fibre varies from 15 per cent to 50 per cent depending on the degree of orientation and nature of crystalline structure within the fibre. The fibre shows moderate recovery from low elongations. The fibre is relatively stiff and possesses excellent resiliency and recovery from bending deformation. The fibre has a specific gravity of 1.38. The fibre is quite hydrophobic, with a moisture regain of 0.1 per cent-0.4 per cent under standard conditions and 1.0 per cent at 21°C and 100 per centRH. It is swollen or dissolved by phenols, chloroacetic acid, or certain chlorinated hydrocarbons at elevated temperatures. The fibre exhibits moderate heat conductivity and has high resistivity, leading to extensive static charge buildup. On heating, the fibre softens in the 210°-250°C range with fibre shrinkage and melts at 250°-255°C.

**Chemical Properties**

Polyethylene terephthalate polyester is highly resistant to chemical attack by acid, bases, oxidising, or reducing agents and is only attacked by hot concentrated acids and bases. The fibre is not attacked by biological agents. On exposure to sunlight, the fibre slowly undergoes oxidative attack without colour change with an accompanying slow loss in strength. The fibre melts at about 250°C with only limited decomposition.

**End-Use Properties**

*Common trade names for polyethylene terephthalate polyester are:*

- Monsanto,
- Dacron,
- Encron,
- Kadel IV,
- Polar Guard,
- Trevira,
- Fortrel, and
- Vycron.

This polyester is an inexpensive fibre with a unique set of desirable properties which have made it useful in a wide range of end-use applications. It has adequate aesthetic properties and bright translucent appearance unless a delusterant has been added to the fibre. The hand of the fibre is somewhat stiff unless the fibre has been texturised or modified, and fabrics from polyester exhibit moderate draping qualities. The fibre is hydrophobic and non-absorbent without chemical modification. This lack of absorbency limits the comfort of polyester fabrics. Polyester possesses good strength and durability characteristics but exhibits moderate to poor recovery from stretching. It has excellent wrinkle resistance and recovery from wrinkling and bending deformation. Polyester shows only fair pilling and snag resistance in most textile constructions.

Heat setting of polyester fabrics provides dimensional stabilisation to the structure, and creases can be set in polyester by heat due to its thermoplastic nature. It has excellent resistance to most household chemicals and is resistant to sunlight-induced oxidative damage, particularly behind window glass. Due to its hydrophobicity and high crystallinity, polyester is difficult to dye, and special dyes and dyeing techniques must be used. When dyed, polyester generally exhibits excellent fastness properties. Polyester has good laundering and dry-cleaning characteristics but is reported to retain oily soil unless treated with appropriate soil release agents.

Owing to its low regain, polyester dries readily and can be safely ironed or dried at temperatures up to 160°C. It is a moderately flammable fibre that burns on contact with a flame but melts and drips and shrinks away from the flame. It has a LOI 21. Polyester is used extensively as a staple and filament fibre in apparel and as a staple fibre in blends with cellulosic fibres for apparel of all types. Polyester is also used extensively in home furnishings and as fibresfill for pillows, sleeping bags, etc. Polyester is used in threads, rope, and tire cord and in sails and nets as well as other industrial fabrics.

## POLY-L,4-CYCLOHEXYLENEDIMETHYLENE TEREPHTHALATE

Poly-l,4-cyclohexylenedimethylene terephthalate resembles polyethylene terephthalate in most properties:

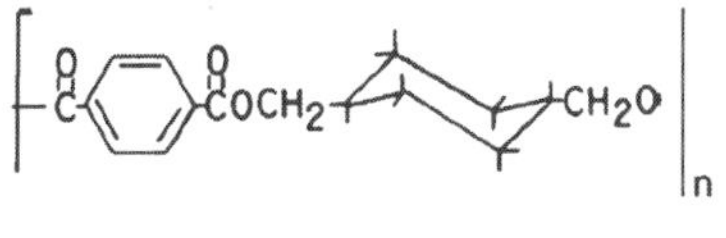

POLY-I.4-CYC LOHEXYLENEDIMETHYL ENE TEREPHTHALATE POLYESTER

The cyclohexylene group within this fibre provides additional rigidity to the molecular chains, but the packing of adjacent polymer chains may be more difficult due to the complex structure. As a result, the fibre has a lower tenacity than polyethylene terephthalate. It has a tenacity of 2.5-3 g/d and exhibits lower elongations than polyethylene terephthalate. It has a lower specific gravity than polyethylene terephthalate.

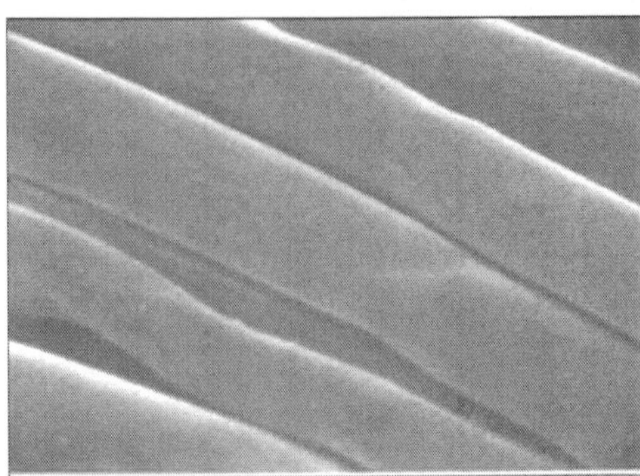

**Fig.** Poly-l,4-cyclohexyLenedimethylene Terephthalate Polyester.

The fibre melts at 290°-295°C and is attacked and shrunk by trichloroethylene and methylene chloride. The fibre has good chemical resistance. The major trade name for this polyester is Kodel II. The fibre is somewhat superior to polyethylene terephthalate in certain end-use properties including better recovery from stretch and better resistance to pillin9. The fibre has superior resiliency and is particularly suited for use in blend with cellulosics and wool, as a carpet fibre, and as fibresfill.

## OTHER POLYESTERS

### Poly-p-ethyleneoxybenzoate

This fibre was introduced by a Japanese firm as A-Tell. The properties of the fibre are very similar to the polyesters containing terephthalate units in most respects including tensile properties, specific gravity, melting point, chemical resistance, and sunlight resistance.

As a result, the FTC expanded the polyester category to include the polymeric £ethyleneoxybenzoate esters. The fibre is produced by reaction of ethylene oxide with £-hydroxybenzoic acid, followed by melt spinning. The fibre has a tenacity of 4-5.5 g/d and an elongation at break of 15 per cent-30 per cent with nearly complete recovery from low elongation. Its specific gravity of 1.34 and moisture regain of 0.4 per cent-5.0 per cent are nearly the same as that of polyethylene terephthalate polyester. The fibre melts at 224°C and softens at about 200°C. The fibre is reported to be even more resistant to attack by acids and bases than terephthalate-based polyesters. The properties of this fibre were not sufficiently different from other polyesters to achieve reasonable market penetration, and the fibre has been discontinued.

### Modified Terephthalate Polyesters

Poor dyeability and the moderate flammabi1ity of polyester have resulted in formulation of modified terephthalate esters to improve the dyeability and the flame retardant properties of the fibre. Introduction of amino or sulfonic acid groups on the benzene ring of terephthalate leads to fibres that are more dyeable with cationic or acid dyes. Bromine, other halogen, or phosphonate groups substituted within the structure provide flame retardant characteristics to the fibre, especially if antimony oxide is present in the fibre matrix. Appropriate non-terephthalate comonomers replacing some of the terephthalate groups introduced into the polyester can improve dyeability and/or flammability of the fibres.

## ACRYLIC FIBRES

Acrylic fibres are synthetic fibres made from a polymer with an average molecular weight of ~100,000, about 1900 monomer units. To be called acrylic in the U.S, the polymer must contain at least 85 per cent acrylonitrile monomer. Typical comonomers are vinyl acetate or methyl acrylate. The Dupont Corporation created the first acrylic fibres in 1941 and trademarked them under the name Orlon.

### PRODUCTION

The polymer is formed by free-radical polymerisation in aqueous suspension. The fibre is produced by dissolving the polymer in a solvent such as N,N-dimethylformamide or aqueous sodium thiocyanate, metering it through a multi-hole spinnerette and coagulating the resultant filaments in an aqueous solution of the same solvent or evaporating the solvent in a stream of heated inert gas. Washing, stretching, drying and crimping complete the processing.

Acrylic fibres are produced in a range of deniers, typically from 0.9 to 15, as cut staple or as a 500,000 to 1 million filament tow. End uses include sweaters, hats, hand-knitting yarns, rugs, awnings, boat covers, and upholstery; the fibre is also used as a precursor for carbon fibre. Production of acrylic fibres is centered in the Far East, Turkey, India, Mexico, and South America, though a number of European producers still continue to operate. All US producers have ended production.

*Former U.S., brands of acrylic were:*

- Acrilan (Monsanto),
- Creslan (American Cyanamid), and
- Orlon (DuPont).

Other brand names that are still in use include Dralon.

### TEXTILE USES

Acrylic is lightweight, soft, and warm, with a wool-like feel. It can also be made to mimic other fibres, such as cotton, when spun on short staple equipment. Some acrylic is extruded in coloured or pigmented form; other is extruded in "ecru", otherwise known as "natural," "raw white," or "undyed." Pigmented fibre has highest light-fastness. Its fibres are very resilient compared to both other synthetics and natural fibres. Some acrylic is used in clothing as a less expensive alternative to cashmere, due to the similar feeling of the materials. Some acrylic fabrics may fuzz or pill easily. Other fibres and fabrics are designed to minimize pilling.

Acrylic takes colour well, is washable, and is generally hypoallergenic. End-uses include socks, hats, gloves, scarves, sweaters, home furnishing fabrics, and awnings. Acrylic is resistant to moths, oils, chemicals, and is very resistant to deterioration from sunlight exposure. However, static and pilling can be a problem in certain fabrications.

Acrylic has a bad reputation amongst some crafters who knit or crochet; acrylic yarn may be perceived as "cheap" because it is typically priced lower than its natural counterparts. The fibre requires heat to "block" or set the shape of the finished garment, and it isn't as warm as alternatives like wool. Some knitters also complain that the fibre "squeaks" when knitted. On the other hand, it can be useful in certain items, like garments for babies, which require constant washing, because it is machine-washable. Acrylic can irritate the skin of people with dermatological conditions such as eczema but this is unusual. Leading US spinners include National Spinning Co., Inc.

*Leading acrylic fibre producers include:*

- AkSa (Turkey),
- Montefibre (Spain),
- Dralon (Germany),
- Kaltex (Mexico), and
- Birla Acrylic (Thailand and Egypt).

## POLYETHYLENE AND POLYPROPYLENE

Polyolefin fibres are those fibres produced from polymers formed by chain growth polymerisation of olefins and which contain greater than 85 per cent polymerised ethylene, propylene, or other olefin units. In general, linear high-density stereoregular polyethylene and polypropylene are used in textile applications, with polypropylene predominating due to its superior temperature stability. These fibres have good strength and toughness, have good abrasion resistance, and are inexpensive. The fibres are difficult to dye and have relatively low melting points, but they are effectively used in a wide variety of textile applications.

### Structural Properties

Linear polyethylene and polypropylene are polymerised from their corresponding monomers to a DP of 1000-3000 using complex metal catalysts at 40°-100°C and at moderate pressures. Using this initiation technique, branching of the polyolefin due to free radical chain transfer is avoided and a linear unbranched structure is formed. With polypropylene, catalyst systems are selected that will lead to regular isotactic placement of the optically active methyl-substituted carbon on the backbone to give a structure capable of high crystallinity:

$-\!(CH_2CH_2)_n\!-$ $-\!(CH_2\underset{CH_3}{CH})_n\!-$

POLYETHYLENE POLYPROPYLENE

ISOTACTIC POLYPROPYLENE

The resultant polymers are dried, compounded with appropriate additives, melt spun into fibres, and drawn to orient. The highly linear chains of these polyolefins can closely pack and associate with adjacent chains through van der Waals interactions and possess crystallinities in the 45 per cent60 per cent range, The surface of these fibres is usually smooth, and the fibre cross section is round. Also, polyolefin films can be split using knife edges to form flat ribbonlike fibres with a rectangular cross section.

**Fig.** Polypropylene.

## Physical Properties

Polypropylene and polyethylene are strong fibres with good elongation and recovery properties. The tenacity of fibres varies from 3.5 to 8 g/d, with an elongation at break of 0 per cent-30 per cent. The fibres recover well from stress, with 95 per cent recovery at 10 per cent elongation. The fibres are moderately stiff and have moderate resiliency on bending. Moisture does not affect these properties, since polyolefins are hydrophobic and have a moisture regain of 0 per cent. The polyethylene fibres have specific gravities of 0.95-0.96 and polypropylene specific gravities of 0.90-0.91. As a consequence these fibres float on water and are the lightest of the major fibres in commerce. The fibres are unaffected by solvents at room temperature and are swollen by aromatic and chlorinated hydrocarbons only at elevated temperatures. They exhibit excellent heat and electric insulation characteristics and are extensively used in these applications. The fibres are heat sensitive. Polyethylene softens at 130°C and melts at 150°C, while polypropylene softens at about 150°C and melts at about 170°C.

## Chemical Properties

The polyolefins are extremely inert and resistant to chemical attack. They are unaffected by chemical and biological agents under normal conditions. They are sensitive to oxidative attack in the pre sence of sunlight due to formation of chromophoric keto groups along the hydrocarbon chain. These groups act as photo sensitisers for further decomposition. The fibre only slowly undergoes oxidative decomposition at its melting point.

**End-Use Properties**

*Common trade names for polyolefin fibres include:*

- Herculon,
- Marvess, and
- Vectra.

Polyolefins are lustrous white translucent fibres with good draping qualities and a characteristic slightly waxy hand. They have excellent abrasion resistance and exhibit fair wrinkle resistance.

They are essentially non-absorbent, but reportedly exhibit wicking action with water. The polyolefins are strong, tough fibres which are chemically inert and resistant to oxidative attack except that induced by sunlight. Stabilising chemicals must be incorporated into the fibres to lower their susceptibility to such light-induced attack.

The polyolefins have high affinity for oil-borne stains, which are difficult to remove on laundering. The fibres are dry-cleanable with normal dry-cleaning solvents if the temperature is kept below SO°C. Since the fibre s have no affinity for water, they dry readily, but care must be taken to keep the drying temperature below 70 °C for polyethylene and 120°C for polypropylene to prevent heat-induced thermoplastic shrinkage and deformation. The fibres are flammable, burn with a black sooty flame, and tend to melt and draw away from the flame.

The polyolefins—particularly polypropylene—have found a number of applications particularly in home furnishing s and industrial fabrics. Uses include indoor-outdoor carpeting, carpet backing, upholstery fabrics, seat covers, webbing for chairs, non-wovens, laundry bags, hosiery and knitwear, fishnet, rope, filters, and industrial fabrics.

# 3

# Fibre Classification

The word "textile" was originally used to define a woven fabric and the processes involved in weaving.

*Over the years the term has taken on broad connotations, including the following:*

- Staple filaments and fibres for use in yarns or preparation of woven, knitted, tufted or non-woven fabrics,
- Yarns made from natural or man-made fibres,
- Fabrics and other products made from fibres or from yarns, and
- Apparel or other articles fabricated from the above which retain the flexibility and drape of the original fabrics.

This broad definition will generally cover all of the products produced by the textile industry intended for intermediate structures or final products. Textile fabrics are planar structures produced by interlacing or entangling yarns or fibres in some manner. In turn, textile yarns are continuous strands made up of textile fibres, the basic physical structures or elements which make up textile products.

Each individual fibre is made up of millions of individual long molecular chains of discrete chemical structure. The arrangement and orientation of these molecules within the individual fibre, as well as the gross cross section and shape of the fibre (morphology), will affect fibre properties, but by far the molecular structure of the long molecular chains which make up the fibre will determine its basic physical and chemical nature. Usually, the polymeric molecular chains found in fibres have a definite chemical sequence which repeats itself along the length of the molecule. The total number of units which repeat themselves in a chain (n) varies from a few units to several hundred and is referred to as the degree of polymerisation (DP) for molecules within that fibre.

## FIBRE CLASSIFICATION

Textile fibres are normally broken down into two main classes, natural and man-made fibres. All fibres which come from natural sources (animals, plants, etc.) and do not require fibre formation or reformation are classed as natural fibres.

Natural fibres include the protein fibres such as wool and silk, the cellulose fibres such as cotton and linen, and the mineral fibre asbestos. Man-made fibres are fibres in which either the basic chemical units have been formed by chemical synthesis followed by fibre formation or the polymers from natural sources have been dissolved and regenerated after passage through a spinneret to form fibres.

Those fibres made by chemical synthesis are often called synthetic fibres, while fibres regenerated from natural polymer sources are called regenerated fibres or natural polymer fibres. In other words, all synthetic fibres and regenerated fibres are man-made fibres, since man is involved in the actual fibre formation process.

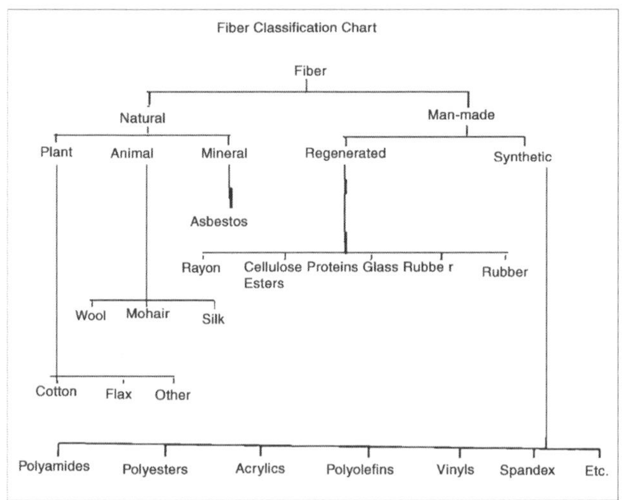


**Fig.** Classification of Natural and Man-made Fibres.

In contrast, fibres from natural sources are provided by nature in ready-made form. The synthetic man-made fibres include the polyamides (nylon), polyesters, acrylics, polyolefins, vinyls, and elastomeric fibres, while the regenerated fibres include rayon, the cellulose acetates, the regenerated proteins, glass and rubber fibres. Figure shows a classification chart for the major fibres.

Another method of classifying fibres would be according to chemical structure without regard of the origin of the fibre and its starting materials. In this manner all fibres of similar chemical structure would be classed together. The natural man-made fibre classification given in figure does this to a certain extent.

In this way, all fibres having the basic cellulosic unit in their structures would be grouped together rather than separated into natural and man-made fibres. This book essentially presents the fibres in groups of similar basic chemical structure, with two exceptions. In one case the elastomeric fibres have been grouped together due to their exceptional physical property, high extensibility and recovery.

## FIBRE PROPERTIES

*There are several primary properties necessary for a polymeric material to make an adequate fibre*:

- Fibre length to width ratio,
- Fibre uniformity,
- Fibre strength and flexibility,
- Fibre extensibility and elasticity, and
- Fibre cohesiveness.

Certain other fibre properties increase its value and desirability in its intended end-use but are not necessary properties essential to make a fibre. Such secondary properties include moisture absorption characteristics, fibre resiliency, abrasion resistance, density, luster, chemical resistance, thermal characteristics, and flammability.

### FIBRE LENGTH TO WIDTH RATIO

Fibrous materials must have sufficient length so that they can be made into twisted yarns. In addition, the width of the fibre (the diameter of the cross section) must be much less than the overall length of the fibre, and usually the fibre diameter should be 1/100 of the length of the fibre. The fibre may be "infinitely" long, as found with continuous filament fibres, or as short as 0.5 inches (1.3 em), as found in staple fibres. Most natural fibres are staple fibres, whereas man-made fibres come in either staple or filament form depending on processing prior to yarn formation.

### FIBRE UNIFORMITY

Fibres suitable for processing into yarns and fabrics must be fairly uniform in shape and size. Without sufficient uniformity of dimensions and properties in a given set of fibres to be twisted into yarn. The actual formation of the yarn may be impossible or the resulting yarn may be weak, rough, and irregular in size and shape and unsuitable for textile usage. Natural fibres must be sorted and graded to assure fibre uniformity, whereas synthetic fibres may be "tailored" by cutting into appropriate uniform 1engths to give a proper degree of fibre uniformity.

### FIBRE STRENGTH AND FLEXIBILITY

A fibre or yarn made from the fibre must possess sufficient strength to be processed into a textile fabric or other textile article. The resulting textile must have sufficient strength to provide adequate durability during end-use. Many experts consider single fibre strength of 5 grams per denier to be necessary for a fibre suitable in most textile applications, although certain fibres with strengths as low as 1.0 gram per denier have been found suitable for some applications.

The strength of a single fibre is called the tenacity, defined as the force per unit linear density necessary to break a known unit of that fibre. The breaking tenacity of a fibre may be expressed in grams per denier (g/d) or grams per tex (g/tex). Both denier and tex are units of 1inear density (mass per unit of fibre length) and are defined as the number of grams of fibre measuring 9000 meters and 1000 meters, respectively. As a result, the denier of a fibre or yarn will always be 9 times the tex of the same fibre. Since tenacities of fibres or yarns are obtained by dividing the force by denier or tex, the tenacity of a fibre in grams per denier will be 1/9 that of the fibre tenacity in grams per tex. As a result of the adaption of the International System of Units. The appropriate length unit for breaking tenacity becomes kilometer (km) of breaking length or Newtons per tex (N/tex) and will be equivalent in value to g/tex.

The strength of a fibre yarn or fabric can be expressed in terms of force per unit area, and when expressed in this way the term is tensile strength. The most common unit used in the past for tensile strength has been pounds force per square inch or grams force per square centimeter. In 51 units, the pounds force per square inch x 6.895 will become kilopascals (kPa) and grams force per square centimeter x 9.807 will become megapascals (MPa). A fibre must be sufficiently flexible to go through repeated bending without significant strength deterioration or breakage of the fibre. Without adequate flexibility, it would be impossible to convert fibres into yarns and fabrics, since flexing and bending of the individual fibres is a necessary part of this conversion. In addition, individual fibres in a textile will be subjected to considerable bending and flexing during enduse.

## FIBRE EXTENSIBILITY AND ELASTICITY

An individual fibre must be able to undergo slight extensions in length (less than 5 per cent) without breakage of the fibre. At the same time the fibre must be able to almost completely recover following slight fibre deformation. In other words, the extension deformation of the fibre must be nearly elastic. These properties are important because the individual fibres in textiles are often subjected to sudden stresses, and the textile must be able to give and recover without significant overall deformation of the textile.

## FIBRE COHESIVENESS

Fibres must be capable of adhering to one another when spun into a yarn. The cohesiveness of the fibre may be due to the shape and contour of the individual fibre s or the nature of the surface of the fibres. In addition, long-filament fibres by virtue of their length can be twisted together to give stability without true cohesion between fibres. Often the term "spinning quality" is used to state the overall attractiveness of fibres for one another.

## MOISTURE ABSORPTION AND DESORPTION

Most fibres tend to absorb moisture (water vapor) when in contact with the atmosphere. The amount of water absorbed by the textile fibre will depend on the chemical and physical structure and properties of the fibre, as well as the temperature and humidity of the surroundings. The percentage absorption of water vapor by a fibre is often expressed as its moisture regain. The regain is determined by weighing a dry fibre, then placing it in a room set to standard temperature and humidity (21° ± 1° C and 65 per cent relative humidity [RH] are commonly used). From these measurements, the percentage moisture regain of the fibre is determined:

$$\text{Percentage Regain} = \frac{\text{Conditioned weight} - \text{Dry Weight}}{\text{Dry Weight}} \times 100\%$$

Percentage moisture content of a fibre is the percentage of the total weight of the fibre which is due to the moisture present, and is obtained from the following formula:

$$\text{Percentage moisture content} = \frac{\text{Conditioned weight} - \text{Dry weight}}{\text{Conditioned weight}} \times 100\%$$

The percentage moisture content will always be the smaller of the two values. Fibres vary greatly in their regain, with hydrophobic (water-repelling) fibres having regains near zero and hydrophilic (water-seeking) fibres like cotton, rayon, and wool having regains as high as 15 per cent at 21°C and 65 per cent RH. The ability of fibres to absorb high regains of water affects the basic properties of the fibre in end-use. Absorbent fibres are able to absorb large amounts of water before they feel wet, an important factor where absorption of perspiration is necessary. Fibres with high regains will be easier to process, finish, and dye in aqueous solutions, but will dry more slowly.

The low regain found for many man-made fibres makes them quick drying, a distinct advantage in certain applications. Fibres with high regains are often desirable because they provide a "breathable" fabric which can conduct moisture from the body to the outside atmosphere readily, due to their favourable moisture absorption-desorption properties. The tensile properties of fibres as well as their dimensional properties are known to be affected by moisture.

## FIBRE RESILIENCY AND ABRASION RESISTANCE

The ability of a fibre to absorb shock and recover from deformation and to be generally resistant to abrasion forces is important to its end-use and wear characteristics. In consumer use, fibres in fabrics are often placed under stress through compression, bending, and twisting (torsion) forces under a variety of temperature and humidity conditions. If the fibres within the fabric possess good elastic recovery properties from such deformative actions, the fibre has good resiliency and better overall appearance in end-use. For example, cotton and wool show poor wrinkle recovery under hot moist conditions, whereas polyester exhibits good recovery from deformation as a result of its high resiliency.

Resistance of a fibre to damage when mobile forces or stresses come in contact with fibre structures is referred to as abrasion resistance. If a fibre is able to effectively absorb and dissipate these forces without damage, the fibre will show good abrasion resistance. The toughness and hardness of the fibre is related to its chemical and physical structure and morphology of the fibre and will influence the abrasion of the fibre. A rigid, brittle fibre such as glass, which is unable to dissipate the forces of abrasive action, results in fibre damage and breakage, whereas a tough but more plastic fibre such as polyester shows better resistance to abrasion forces. Finishes can affect fibre properties including resiliency and abrasion resistance.

## LUSTER

Luster refers to the degree of light that is reflected from the surface of *a* fibre or the degree of gloss or sheen that the fibre possesses. The inherent chemical and physical structure and shape of the possesses. The inherent chemical and physical structure and shape of the fibre *can* affect the relative luster of the fibre. With natural fibres the luster of the fibre is dependent on the morphological form that nature gives the fibre, although the relative luster can be changed by chemical and/or physical treatment of the fibre as found in processes such *as* mercerisation of cotton.

Man-made fibres can vary in luster from bright to dull depending on the amount of delusterant added to the fibre. Oelusterants such *as* titanium dioxide tend to scatter and absorb light, thereby making the fibre appear duller. The desirability of luster for *a* given fibre application will vary and is often dependent on the intended end-use of the fibre in *a* fabric or garment form and on current fashion trends.

## RESISTANCE TO CHEMICALS IN THE ENVIRONMENT

A textile fibre to be useful must have reasonable resistance to chemicals it comes in contact with in its environment during use and maintenance. It should have resistance to oxidation by oxygen and other gases in the air, particularly in the presence of light, and be resistant to attack by microorganisms and other biological agents. Many fibres undergo light-induced reactions, and fibres from natural sources are susceptible to biological attack, but such deficiencies can be minimized by treatment with appropriate finishes. Textile fibres come in contact with a large range of chemical agents on laundering and dry cleaning and must be resistant from attack under such conditions.

## OENSITY

The density of a fibre is related to its inherent chemical structure and the packing of the molecular chains within that structure. The density of a fibre will have a noticeable effect on its aesthetic appeal and its usefulness in given applications.

For example, glass and silk fabrics of the same denier would have noticeable differences in weight due to their broad differences in density. Fishnets of polypropylene fibres are of great utility because their density is less than that of water. Oensities are usually expressed in units of grams per cubic centimeter, but in 51 units will be expressed as kilograms per cubic meter, which gives a value 1000 times larger.

## THERMAL AND FLAMMABILITY CHARACTERISTICS

Fibres used in textiles must be resistant to wet and dry heat, must not ignite readily when coming in contact with a flame, and ideally should self-extinguish when the flame is removed. Heat stability is particularly important to a fibre during dyeing and finishing of the textile and during cleaning and general maintenance by the consumer.

Textile fibres for the most part are made up of organic polymeric materials containing carbon and burn on ignition from a flame or other propagating source. The chemical structure of a fibre establishes its overall flammability characteristics, and appropriate textile finishes can reduce the degree of flammability.

A number of Federal, state, and local statutes eliminate the most dangerous flammable fabrics from the marketplace.

## PRIMARY FIBRE PROPERTIES FROM AN ENGINEERING PERSPECTIVE

The textile and polymer engineer must consider a number of criteria essential for formation, fabrication, and assembly of fibres into textile substrates. Often the criteria used will be similar to those set forth above concerning end-use properties.

*Ideally a textile fibre should have the following properties:*

- A melting and/or decomposition point above 220°C.
- A tensile strength of 5 g/denier or greater.
- Elongation at break above 10 per cent with reversible elongation up to 5 per cent strain.
- A moisture absorptivity of 2 per cent-5 per cent moisture uptake.
- Combined moisture regain and air entrapment capability.
- High abrasion resistance.
- Resistance to attack, swelling, or solution in solvents, acids, and bases.
- Self-extinguishing when removed from a flame.

## FIBRE FORMATION AND MORPHOLOGY

Fibre morphology refers to the form and structure of a fibre, including the molecular arrangement of individual molecules and groups of molecules within the fibre.

Most fibres are organic materials derived from carbon combined with other atoms such as oxygen, nitrogen, and halogens. The basic building blocks that organic materials form as covalently-bonded organic compounds are called monomers.

Covalent bonds involve the sharing of electrons between adjacent atoms within the monomer, and the structure tons between adjacent atoms within the monomer, and the structure of the monomer is determined by the type, location, and nature of bonding of atoms within the monomer and by the nature of covalent bonding between atoms.

Monomers react or condense to form long-chain molecules called polymers made up of a given number (n) of monomer units which are the basic building unit of fibres.

On formation into fibres and orientation by natural or mechanical means the polymeric molecules possess ordered crystalline and non-ordered amorphous areas, depending on the nature of the polymer and the relative packing of molecules within the fibre.

For a monomer A the sequence of events to fibre formation and orientation would appear as shown in figure.

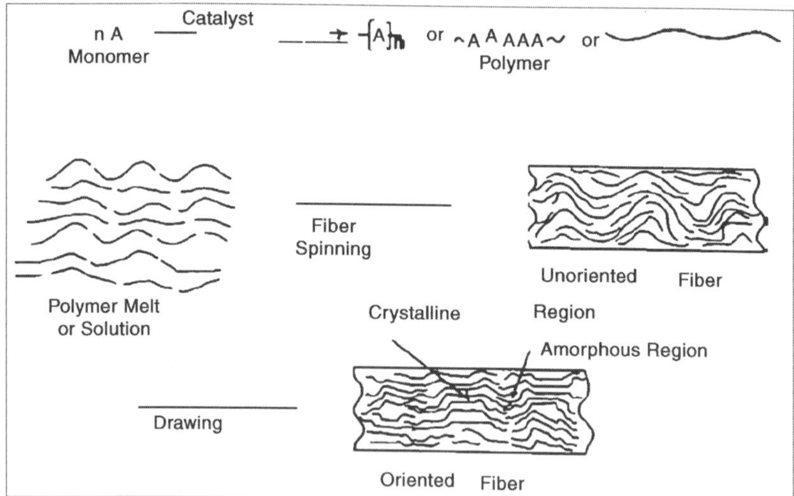


**Fig.** Polymerisation Sequence and Fibre Formation.

Polymers with repeating units of the same monomer (An) would be referred to as homopolymers. If a second unit B is introduced into the basic structure, structures which are copolymers are formed with structures as outlined in figure.

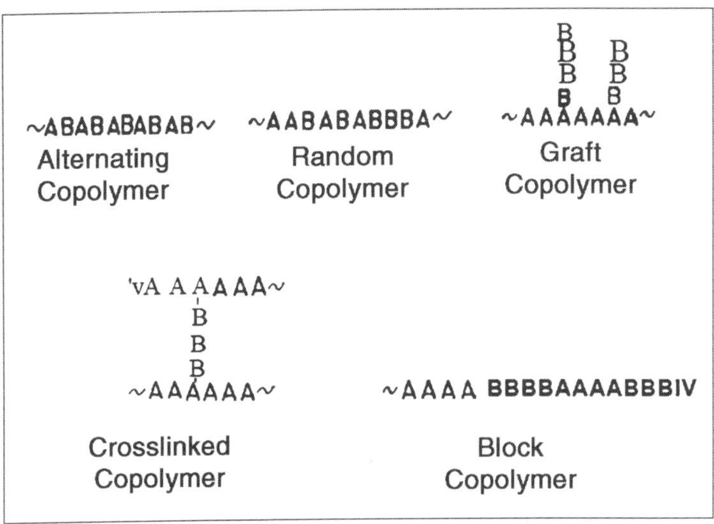


**Fig.** Copolymer Structures.

## POLYMER FORMATION

Synthetic polymers used to form fibres are often classified on the basis of their mechanism of polymerisation—step growth (condensation) or chain growth (addition) polymerisation.

Step growth polymerisation involves multifunctional monomers which undergo successive condensation with a second monomer or with itself to form a dimer, which in turn condenses with another dimer to form a tetramer, etc., usually with loss of a small molecule such as water. Chain growth involves the instantaneous growth of a long molecular chain from unsaturated monomer units, followed by initiation of a second chain, etc. The two methods are outlined schematically:

$$\text{Step Growth:} \quad nA \rightarrow \frac{n}{2}AA \rightarrow AAAA \rightarrow$$

$$\text{Chain Growth:} \quad nA \rightarrow (A)n \rightarrow (A)n \qquad nA \rightarrow (A)n$$

The average number of monomer repeating units in a polymer chain (n) is often referred to also as the degree of polymerisation, DP. The DP must exceed an average 20 units in most cases to give a polymer of sufficient molecular size to have desirable fibre-forming properties. The overall breadth of distribution of molecular chain lengths in the polymer will affect the ultimate properties of the fibres, with wide polymer size distributions leading to an overall reduction of fibre properties. Although the polymers from natural fibres and regenerated natural fibres do not undergo polymerisation by the mechanisms found for synthetic fibres, most natural polymers have characteristic repeating units and high degrees of polymerisation and are related to step growth polymers. Basic polymeric structures for the major fibres are given in figure.

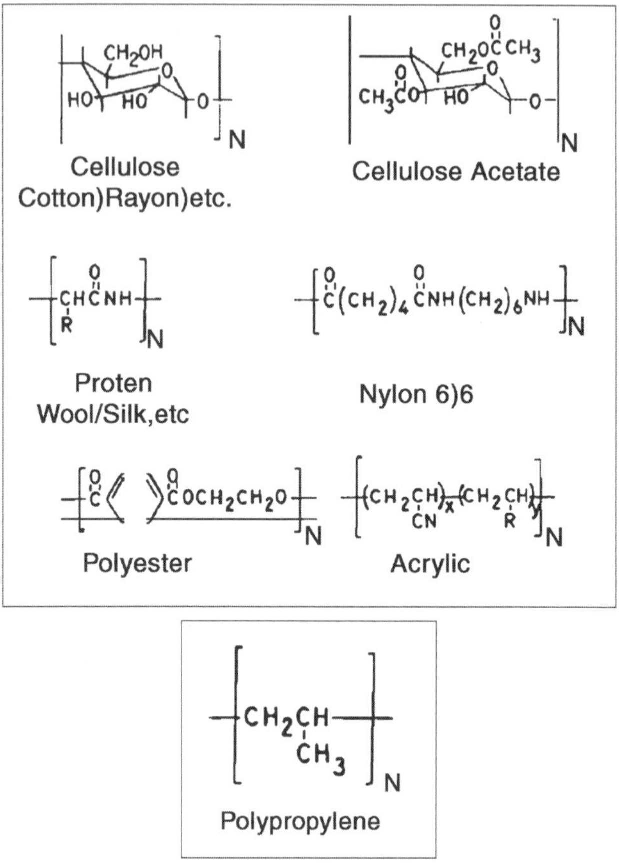

Cellulose
Cotton)Rayon)etc.

Cellulose Acetate

Proten
Wool/Silk,etc

Nylon 6)6

Polyester

Acrylic

Polypropylene

**Fig.** Basic Polymeric Structures for Major Fibres.

## FIBRE SPINNING

Although natural fibres come in a morphological form determined by nature, regenerated and synthetic man-made fibres can be "tailor-made" depending on the shape and dimensions of the orifice (spinning jet) that the polymer is forced through to form the fibre. There are several methods used to spin a fibre from its polymer, including melt, dry, wet, emulsion, and suspension spinning.

Melt spinning is the least complex of the methods. The polymer from which the fibre is made is melted and then forced through a spinneret and into air to cause solidification and fibre formation. Dry and wet spinning processes involve dissolving the fibre-forming polymer in an appropriate solvent, followed by passing a concentrated solution (20 per cent- 50 per cent polymer) through the spinneret and into dry air to evaporate the solvent in the case of dry spinning or into a coagulating bath to cause precipitation or regeneration of the polymer in fibre form in the case of wet spinning.

There is a net contraction of the spinning solution on loss of solvent. If a skin of polymer is formed on the fibre followed by diffusion of the remainder of the solvent from the core of the forming fibre, the cross-section of the fibre as it contracts may collapse to form an irregular popcorn like cross section. Emulsion spinning is used only for those fibre-forming polymers that are insoluble. Polymer is mixed with a surface-active agent (detergent) and possibly a solvent and then mixed at high speeds with water to form an emulsion of the polymer. The polymer is passed through the spinneret and into a coagulating bath to form the fibre. In suspension spinning, the polymer is swollen and suspended in a swelling solvent. The swollen suspended polymer is forced through the spinneret into dry hot air to drive off solvent or into a wet non-solvent bath to cause the fibre to form through coagulation.

*The spinning process can be divided into three steps:*

1. Flow of spinning fluid within and through the spinneret under high stress and sheer.
2. Exit of fluid from the spinneret with relief of stress and an increase in volume (ballooning of flow).
3. Elongation of the fluid jet as it is subjected to tensile force as it cools and solidifies with orientation of molecular structure within the fibre.

*Common cross sections of man-made fibres include:*

- Round,
- Trilobal,
- Pentalobal,
- Dog-bone, and
- Crescent shapes.

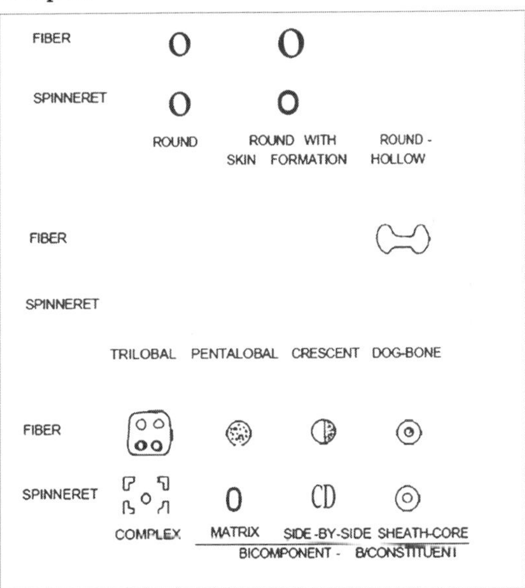


**Fig.** Fibre Cross Sections.

When two polymers are used in fibre formation as in bicomponent or biconstituent fibres, the two components can be arranged in a matrix, side-by-side, or sheath-core configuration. Round cross sections are also found where skin formation has caused fibre contraction and puckering (as with rayons) has occurred or where the spinneret shape has provided a hollow fibre. Complex fibre cross-sectional shapes with special properties are also used.

## FIBRE DRAWING AND MORPHOLOGY

On drawing and orientation the man-made fibres become smaller in diameter and more crystalline, and imperfections in the fibre morphology are improved somewhat. Side-by-side bi component or bi constituent fibres on drawing become wavy and bulky.

In natural fibres the orientation of the molecules within the fibre is determined by the biological source during the growth and maturity process of the fibre. The form and structure of polymer molecules with relation to each other within the fibre will depend on the relative alignment of the molecules in relationship to one another. Those areas where the polymer chains are closely aligned and packed close together are crystalline areas within the fibre, whereas those areas where there is essentially no molecular alignment are referred to as amorphous areas. Dyes and finishes can penetrate the amorphous portion of the fibre, but not the ordered crystalline portion.

## CROSS-SECTION

A number of theories exist concerning the arrangement of crystalline and amorphous areas within a fibre. Individual crystal line areas in a fibre are often referred to as microfibrils. Microfibrils can associate into larger crystalline groups, which are called fibrils or micelles. Microfibrils are 30-100 Å in length, whereas fibrils and micelles are usually 200-600 Å in length. This compares to the individual molecular chains, which vary from 300 to 1500 Å in length and which are usually part of both crystal line and amorphous areas of the polymer and therefore give continuity and association of the various crystal line and amorphous areas within the fibre.

A number of theories have been developed to explain the interconnection of crystal line and amorphous areas in the fibre and include such concepts as fringed micelles or fringed fibrils, molecular chain folding, and extended chain concepts. The amorphous areas within a fibre will be relatively loosely packed and associated with each other, and spaces or voids will appear due to discontinuities within the structure. Figure outlines the various aspects of internal fibre morphology with regard to polymer chains.

The force s that keep crystalline areas together within a fibre include chemical bonds (covalent, ionic ) as well as secondary bonds. Cova lent bonds result from sharing of electrons between atoms, such as found in carboncarbon,

carbon-oxygen, and carbon-nitrogen bonding, within organic compounds. Covalent bonds joining adjacent polymer chains are referred to as crosslinks. Ionic bonding occurs when molecules donate or accept electrons from each other, a s when a metal salt reacts with acid side chains on a po1ymer within a fibre.

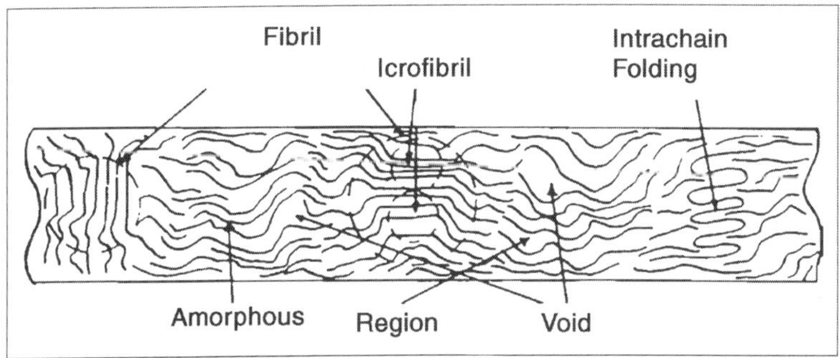


**Fig.** Aspects of Internal Fibre Morphology.

Chemical bonds a re much stronger than secondary bonds formed between polymer chains, but the total associative force between polymer chains can be large, since a very large number of such bonds may occur between adjacent polymer chains. Hydrogen bonds are the strongest of the secondary bonds and occur between electropositive hydrogen atoms and electronegative atoms such as oxygen, nitrogen, and halogen s on opposing polymer chains. Nylon, protein, and cellulosic fibres are capable of extensive hydrogen bonding. Van der Waals interactions between polymer chains occur when clouds of electrons from each chain come in close proximity, thereby promoting a small attractive force between chains.

The more extended the cloud of electrons, the stronger the van der Waals interaction will be. Covalent bonded materials will show some uneven distribution of electron density over the molecule due to the differing electronegativity of the atoms and electron distribution over the molecule to form dipoles.

Dipoles on adjacent polymer chains of opposite charge and close proximity are attracted to each other and promote secondary bonding. When a synthetic fibre is stretched or drawn, the molecules in mo s t cases will orient themselves in crystalline areas parallel to the fibre axis, although crystalline areas in some chain-folded polymers such as polypropylene can be aligned vertical to the fibre axis. The degree of crystallinity will be affected by the total forces available for chain interaction, the distance between parallel chains, and the similarity and uniformity of adjacent chains.

The structure and arrangement of individual polymer chains also affects the morphology of the fibre. Also, configurations or optical isomers of polymers can have very different physical and chemical properties.

## BULKING, TEXTURISING, AND STAPLE FORMATION

Thermoplastic man-made fibre s can be permanently heat-set after drawing and or ientation.

The fibre will possess structural integrity and will not shrink up to that setting temperature.

*Al so, thermoplastic fibres or yarns from this fibre s can be texturised to give three dimensional loft and bulkiness:*

1. Through fibre deformation and setting at or near their softening temperature,
2. Through air entanglement, or
3. Through differential setting within fibres or yarns.

**Table. Texturising Methods.**

| Heat-setting Techniques | Air Entanglement | Differential Setting |
|---|---|---|
| False Twist | Air Jet | Bicomoponent- |
| Knife edge | | biconstituent |
| Stuffer box | | Fibre orientation |
| Gear crimping | | Heat shrinkage of |
| Autowist | | thermoplastic fibres |
| Knit-de-knit | | in a blend |

Schematic representations of these methods are given in figure.

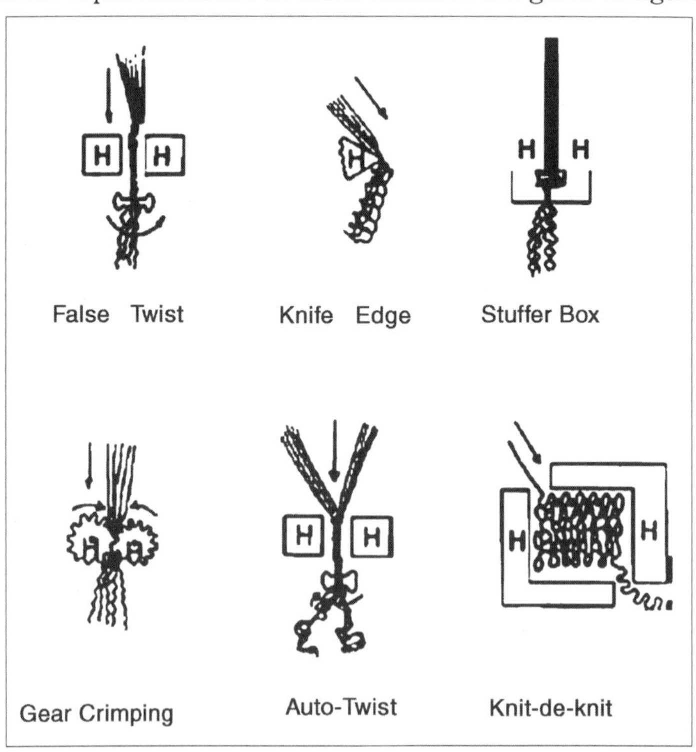

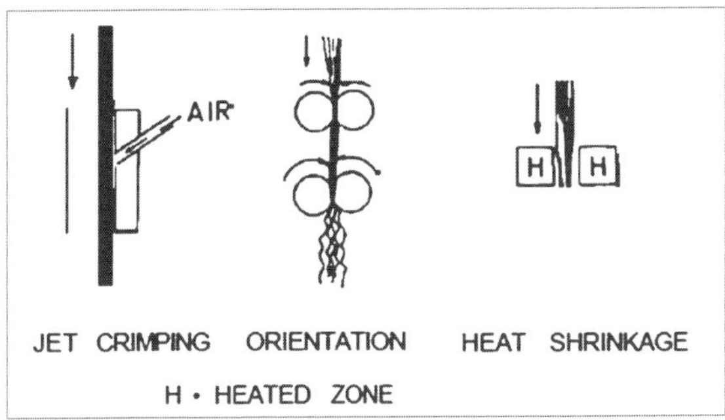


**Fig.** Texturising Methods.

## Heat-Setting Techniques

The false twist heat-setting technique is extremely rapid, inexpensive, and most widely used. The filament fibre tow is brought in contact with a high-speed spindle running vertical to the moving tow. This action results in a high twist in the tow up to the spindle. The twisted tow is heated near its softening point before passing the spindle and then cooled and untwisted to give a wavy bulky yarn.

In knife edge texturising, filament tows or yarns are passed over a heated knife edge under tension. The fibres near the knife edge are changed in overall orientation in relation to the unheated yarns or portion of the filaments away from the knife edge, thereby causing bulking of the yarn. In stuffer box texturising, the filament tow is fed into a heated box, causing the tow to double up against itself. On removal, the cooled tow retains the zigzag configuration caused by the process. In gear crimping, the tow is passed between heated intermeshing gears. On cooling the fibres retain the shape induced by the heated gears. In auto-twisting, two tows or yarns are twisted together and then heat set. On untwisting the yarns have equal but opposite twists, which causes a spiral bulking of the yarn. In the knit-de-knit process a yarn is fill knitted, heat set, cooled, and deknitted to give a bulked yarn retaining the shape and curvature of the knit.

## Air Entanglement

In air entanglement texturising, a fibre tow is loosely fed into and through a restricted space and a high-speed air jet is impinged on the fibres at a *45°* angle. The loose fibres within the tow are looped to give a texturised effect.

## Differential Setting

Heat shrinkage techniques cause a bulking of fibre tows containing different fibres through heating one component of the blend sufficiently to cause heat shrinkage of the fibre and compaction, contraction, and bulking. Side-by-side bicomponent and biconstituent fibres recover different degrees on each side from fibre stretching and causing a waving, crimping, or bulking of the fibre.

## STAPLE FORMATION

Continuous filaments can be cut into staple by wet or dry cutting techniques. In wet cutting, the wet spun fibre is cut to uniform lengths right after spinning, while dry cutting involves partial cutting, debonding, and shuffling of the dry tow to form a sliver. Before the filament or staple is used in yarn spinning, spin finishes are added to give lubricity and antistatic characteristics to the fibres and to provide a greater degree of fibre cohesiveness. Such finishes are usually mixtures including such materials as fatty acid esters, mineral oils, synthetic esters, silicones, cationic amines, phosphate esters, emulsifiers, and/or non-ionic surfactants. Spin finishes are formulated to be oxidation resistant, to be easily removed by scouring, to give a controlled viscosity, to be stable to corrosion, to resist odour and colour formation, and to be non-volatile and readily emulsifiable.

## STRUCTURE-PROPERTY RELATIONSHIPS

The basic chemical and morphological structure of polymers in a fibre determine the fundamental properties of a fabric made from that fibre. Although physical and chemical treatments and changes in yarn and fabric formation parameters can alter the fabric properties to some degree, the basic properties of the fabric result from physical and chemical properties inherent to the structure of the polymer making up the fibre. From these basic properties, the end-use characteristics of the fibre are determined. To that end, in the following stages as suggested, describe the various textile fibres in terms of their basic structural properties, followed by physical and chemical properties, and finally the end-use characteristics inherent to constructions made from the fibre. Initially the name and general information for a given fibre is set forth followed by an outline of the structural properties, including information about chemical structure of the polymer, degree of polymerisation, and arrangement of molecular chains within the fibre.

Physical properties include mechanical (tensile) and environmental properties of the fibre, whereas the effect of common chemicals and chemically induced processes on the fibre is listed under chemical properties. The end-use properties are then listed and include properties coming inherently from the structural, physical, and chemical properties of the fibre as well as end-use properties that involve evaluation of performance, subjective aspects, and aesthetics of the fabrics. Where possible, the interrelationships of these properties are presented.

# CELLULOSIC FIBRES

Cellulose is a fibrous material of plant Origin and the basis of all natural and man-made cellulosic fibres. The natural cellulosic fibres include cotton, flax, hemp, jute, and ramie. The major man-made cellulosic fibre is rayon, a fibre produced by regeneration of dissolved forms of cellulose. Cellulose is a polymeric sugar made up of repeating 1,4-8-anhydroglucose units connected to each other by 8-ether linkages.

The number of repeating units in cellulosic fibres can vary from less than 1000 to as many as 18,000, depending on the fibre source. Cellulose is,a hemiacetal and hydrolyses in dilute acid solutions to form glucose, a simple sugar. The long 1inear chains of cellulose permit the hydroxyl functional groups on each anhydroglucose unit to interact with hydroxyl groups on adjacent chains through hydrogen bonding and van der Waal s forces. These strong intermolecular forces between chains, coupled with the high linearity of the cellulose molecule, account for the crystalline nature of cellulosic fibres.

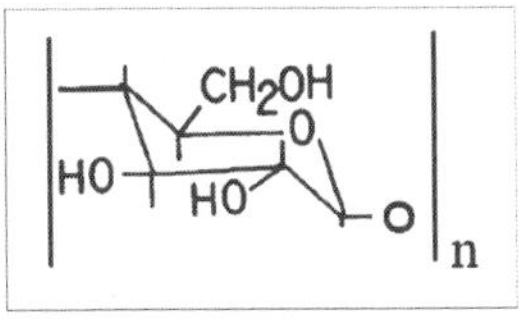

**Fig.** Cellulose.

It is believed that a gradual transition from alternating areas of greater molecular alignment or crystallinity to more disordered or amorphous areas occurs in cellulose. The number, size, and arrangement of crystalline regions within celluloses determine the ultimate properties of a particular fibre. The predominant reactive groups within cellulose are the primary and secondary hydroxyl functional groups.

Each repeating anhydroglucose unit contains one primary and two secondary hydroxyl functional groups which are capable of undergoing characteristic chemical reactions of hydroxyl groups. The primary hydroxyls are more accessible and reactive than secondary hydroxyls; nevertheless, both types enter into many of the chemical reactions characteristic of cellulose.

## COTTON

Cotton is the most important of the natural cellulosic fibres. It still accounts for about 50 per cent of the total fibre production of the world, although man-made fibres have made significant inroads into cotton's share during the last three decades. Cotton fibres grow in the seed hair pod of cotton plants grown and cultivated in warm climates.

### Structural Properties

Cotton is very nearly pure cellulose. As many as 10,000 repeating anhydroglucose units are found in the polymeric cellulosic chains of cotton. Studies have shown that all of the hydroxyl hydrogens in cotton are hydrogen bonded. These hydrogen bonds will hold several adjacent cellulose chains in close alignment to form crystalline areas called microfibrils. These microfibrils in turn align themselves with each other to form larger crystalline units called fibrils, which are visible under the electron microscope.

In cotton the fibrils are laid down in a spiral fashion within the fibre. Modern fibre theory suggests that each cellulose molecule is present within two or more crystalline regions of cellulose will be held together. Between the crystalline regions in cotton, amorphous unordered regions are found. Voids, spaces, and irregularities in structure will occur in these amorphous areas, whereas the cellulose chains in crystalline regions will be tightly packed. Penetration of dyestuffs and chemicals occurs more readily in these amorphous regions.

Approximately 70 per cent of the cotton fibre is crystalline. Individual cotton fibres are ribbonlike structures of somewhat irregular diameter with periodic twists or convolutions along the length of the fibre. These characteristic convolutions, as well as the cross-sectional shape of cotton are caused by collapse of the mature fibre on drying.

**Fig.** Cotton.

Three basic areas exist within the cross section of a cotton fibre. The primary outer wall or cuticle of cotton is a protective tough shell for the fibre, while the secondary wall beneath the outer shell makes up the bulk of the fibre. The fibrils within the secondary wall are packed alongside each other aligned as spiral s running the length of the fibre. The lumen in the center of the fibre is a narrow canal-like structure running the length of the fibre. The lumen carries nutrients to the fibre during growth, but on maturity the fibre dries and the lumen collapses.

### Physical Properties

The high crystallinity and associative forces between chains in cotton result in a moderately strong fibre having a tenacity of 2-5 g/d. The hydrophilic nature of cotton and the effect of absorbed water on the hydrogen bonding within cotton cause the tensile strength of cotton to change significantly with changes in moisture content. As a result, wet cotton is about 20 per cent stronger than dry cotton. Cotton breaks at elongations of less than 10 per cent, and the elastic recovery of cotton is only 75 per cent after only 2 per cent elongation.

Cotton is a relatively stiff fibre; however, wetting of the fibre with water plasticises the cellulose structure, and the cotton becomes more pliable and soft. The resiliency of dry and wet cotton is poor, and many finishes have been developed to improve the wrinkle recovery characteristics of cotton. Cotton is one of the more dense fibres and has a specific gravity of 1.54. The hydroxyl groups of cotton possess great affinity for water, and the moisture regain of cotton is 7 per cent- 9 per cent under standard conditions. At 100 per cent relative humidity, cotton has 25 per cent-30 per cent moisture absorbency.

The heat conductivity of cotton is high, and cotton fabrics feel cool to the touch. Cotton has excellent heat characteristics, and its physical properties are unchanged by heating at 120°C for moderate periods. The electrical resistivity of cotton is low at moderate relative humidities, and the fibre has low static electricity buildup characteristics. Cotton is not dissolved by common organic solvents. Cotton is swollen slightly by water because of its hydrophilic nature, but it is soluble only in solvents capable of breaking down the associative forces within the crystalline areas of cotton. Aqueous cuprammonium hydroxide and cupriethylenediamine are such solvents.

**Chemical Properties**

Cotton is hydrolysed by hot dilute or cold concentrated acids to form hydrocellulose but is not affected by dilute acids near room temperature. Cotton has excellent resistance to alkalies. Concentrated alkali solutions swell cotton, but the fibre is not damaged. The swelling of cotton by concentrated sodium hydroxide solution is used to chemically finish cotton by a technique called mercerisation. In aqueous alkali, the cotton swells to form a more circular cross section and at the same time loses convolutions.

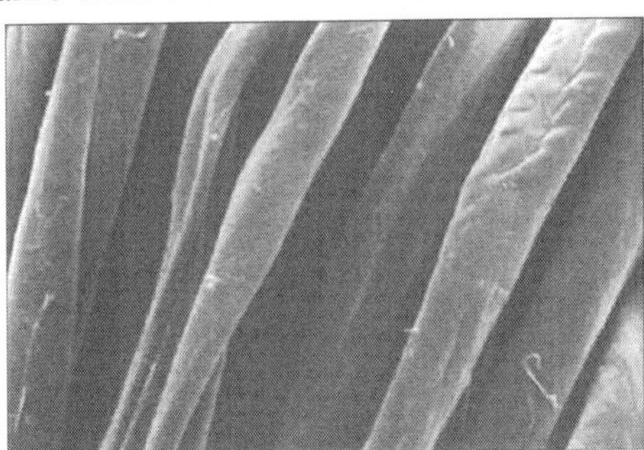

**Fig.** Mercerised Cotton.

If the cotton is held fast during swelling to prevent shrinkage, the cellulose fibers deform to give a fiber of smoother surface. After washing to remove alkali, followed by drying, the cotton fiber retain s a more cylindrical shape and circular cross section. Although little chemical difference exists between mercerized and unmercerized cotton, mercerization does give a more reactive fiber with a higher regain and better dyeability.

Dilute solutions of oxidising and reducing agents have little effect on cotton; however, appreciable attack by concentrated solutions of hydrogen peroxide, sodium chlorite, and sodium hypochlorite is found. Most insects do not attack cotton; however, silverfish will attack cotton in the presence of starch. A major problem with cotton results from fungi and bacteria being able to grow on cotton.

Mildews feed on hot moist cotton fibres, causing rotting and weakening of the fibres. Characteristic odour and pigment staining of the cotton occurs when mildews attack. Additives capable of protecting cotton are available and commercially applied to cotton fabrics used outdoors. These materials are often metal salts of organic compounds which are capable of inhibiting growth of mildews and similar organisms. Cotton is only slowly attacked by sunlight, since cellulose lacks for the most part groups which absorb ultraviolet light between 300 and 400 nm.

Over long periods sunlight degrades cotton, causing it to lose strength and to turn yellow. Certain vat dyes tend to accelerate the rate of cotton photo degradation through a sensitisation reaction called "photo tendering." Although cotton has excellent heat resistance, degradation due to oxidation becomes noticeable when cotton is heated in the air at I50°C for long periods. Spontaneous ignition and burning of cotton occurs at 390°C. At low humidities in the absence of heat and light, cotton will not deteriorate over long periods of storage.

**End-Use Properties**

The properties of cotton fibre are such that it serves as nature's utility fibre. Although cotton has some properties which are undesirable from the viewpoint of the consumer, the superior properties of cotton, coupled with its low cost, nevertheless make it a valuable fibre in many applications. Different species of cotton produce fibres of various average lengths.

*In the United States lengths of cotton staples are designated as follows:*

- Extra long staple: 1 3/8 - 2 inches (3.2 - 4.7 em)
- Long staple: 1 1/8 - 1 3/8 inches (2. 7 - 3. 2 em)
- Medium staple: 1 - 1 1/8 inches (2.5 - 2.7 em)
- Short staple: 7/8 - 1 inch (2.0 - 2.5 em)

The two major types of American cotton are American-Upland and American-Egyptian. Cotton has excellent hand, and the drivability of cotton fabrics is quite acceptable. Fabrics of cotton are of sat is factory appearance and have a low luster unless mercerised or resin finished. The superior absorbency of cotton, coupled with its ability to desorb moisture, makes it a very comfortable fibre to wear. This absorbency permits cotton to be used in applications where moisture absorption is important, such as in sheets and towels. Mildewing of cotton under hot moist conditions and its slowness of drying are undesirable properties associated with its high affinity for water.

The cotton fibre has sufficient strength in the dry and wet states to make it suitable for most consumer textile applications. The increased strength of cotton on wetting adds to its long useful life. Cotton wears well without undue abrasion, and pills do not tend to form as it wears. Cotton's low resiliency and poor recovery from deformation means that it wrinkles easily in both the dry and wet states and exhibits inferior crease retention. Starching of cotton improves these properties, but the effect is only temporary, and it is necessary to renew this finish after each laundering. The resistance of cotton to common household chemicals, sunlight, and heat makes it durable in most textile applications.

Cotton can be dyed successfully by a wide variety of dyes, and the colourfastness of properly dyed cotton is satisfactory. Fabrics of cotton are maintained with a moderate degree of care. Cotton fabrics launder readily, and its wet strength and alkali resistance mean that cotton is resistant to repeated washings. Stresses which occur during the spinning and weaving process will cause cotton fabrics to undergo relaxation shrinkage during initial launderings.

Relaxation shrinkage can be controlled through resin stabilisation or through the well known compression shrinkage process called Sanforisation. Cotton can be drip dried or tumbled dry, but in both cases the dry cotton will be severely wrinkled, and ironing will be necessary. Cotton can be ironed safely at temperatures as high as 205 °C. Cotton, as well as all cellulosic fibres, is highly flammable and continues to burn after removal from a flame. After extinguishing of the flame the cotton will continue to glow and oxidise by a smoldering process called afterglow.

The Limiting Oxygen Index of cotton is 18. A number of topical treatments have been developed to lower the flammability of cot ton and other cellulosics. The undesirable properties of cotton can be corrected to varying degrees through treatment of the fibre with special finishes; however, the abrasion resistance of cotton is adversely affected. As a result, blends of cotton with the stronger man-made fibres have become important. Although man-made fibres have made inroads into applications previously reserved for cotton, cotton continues to be the major textile fibre due to its great versatility, availability, and cost.

## FLAX

Flax is a bats fibre used to manufacture linen textiles. It is derived from the stem of the annual plant, Linumusitatissimum, which grows in many temperate and subtropical areas of the world. The flax plant grows as high as 4 feet tall, and flax fibres are found below the plant surface held in place by woody material and cellular matter.

The flax fibres are freed from thc plant by a fermentation process called retting. Chemical retting using acids and bases has been employed with some success; however, such processes tend to be more expensive than natural fermentation techniques. The fibres are then removed from the plant through breaking of the woody core, removal of the woody material, and combing. The resulting fibres are ready for spinning and are 12-15 inches (30-38 cm) in length.

**Structural Properties**

Flax is nearly pure cellulose and therefore exhibits many properties similar to cotton. Strands of individual flax fibres may consist of many individual fibre cells—fibrils held together by a natural cellulosic adhesive material. The molecular chains in flax are extremely long, and the average molecular weight of molecules in flax is 3 million. There are swellings or nodes periodically along the fibre which show up as characteristic cross markings. The cell walls of flax are thick, and the fibre cross section is polygonal with a large lumen in the center.

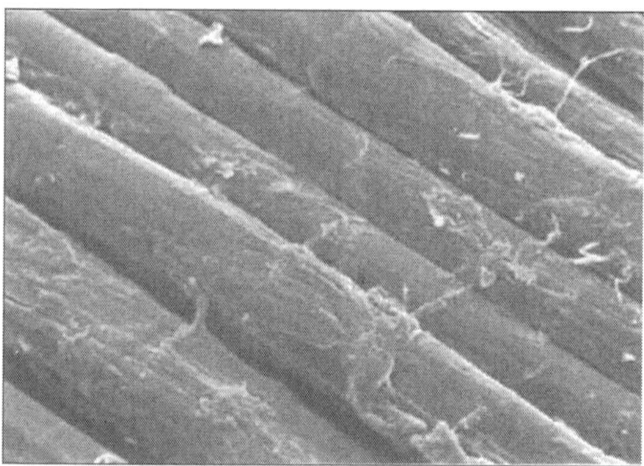

**Fig.** Flax.

**Physical Properties**

Flax is a strong fibre, and the average tenacity of dry flax is 2-7 g/d; wet flax is 2.5-9 g/d. Like cotton, the wet flax fibre is about 20 per cent stronger than the dry fibre.

Flax has a low elongation at break; however, the fibre is fairly elastic at very low elongations. Flax has a higher regain than cotton, and a 12 per cent regain is observed at standard conditions. Flax possesses good heat conductivity and is cool to the touch.

The heat and electrical properties of flax are similar to those of cotton, and flax possesses solubility characteristics identical to cotton. Flax is a highly rigid fibre, and it tends to crease on bending due to its poor resiliency. The specific gravity of flax is 1.54, the same as cotton. Flax is a dull fibre, but more lustrous linen fabrics can be obtained by pounding the linen with wooden hammers. Linen fabrics have good dimensional stability unless stresses have been introduced during the weaving process.

### Chemical Properties

The chemical properties of flax are similar to those of cotton. Flax has good resistance to acids, bases, and chemical bleaches. Flax is resistant to insects and microorganisms, and only under severe moist warm conditions will it be attacked by mildews. Flax is only slowly degraded by sunlight, and it decomposes at temperatures similar to that observed for cotton.

### End-Use Properties

Flax finds only limited use in modern textiles. The methods used to produce flax involve extensive labour and cost; thus flax usage is reserved for special luxury applications. The strength of flax makes it superior to cotton in certain applications. Flax is resistant to abrasion, but like cotton suffers from lack of crease retention and poor wrinkle recovery. Linen fabric's superior absorbency, coupled with its cool crisp hand, contributes to its desirability as a prestige fibre. The thermal properties and chemical resistance of flax make it suitable for many consumer applications. Linen can be laundered repeatedly without deterioration and ironed safely at temperatures as high as 220°C. Flax is very flammable. Flax is used in prestige items including handkerchiefs, towels, tablecloths, sheets, and certain garments.

## OTHER NATURAL CELLULOSIC FIBRES

A large number of natural cellulosic fibres from plant sources exist, including jute, hemp, ramie, kenaf, urena, sisal, henequen, abaca, pina, kapok, coir, and others. Each of the fibres is a plant or seed hair fibre and has properties that make it suitable for selected applications. The major fibres—jute, hemp, and ramie—will be considered here. Hemp is a bast fibre harvested from the hemp plant and processed in a manner similar to flax. It is a coarser fibre than flax, darker in colour and difficult to bleach. The fibre is strong and durable, and the strands of hemp fibre may reach six feet or longer. Individual hemp cells are 1/2-1 inch in length, and the fibre cross section is polygonal. The fibre is very stiff and contains considerable lignin. Although fine fabrics can be produced from select hemps, hemp is used mainly in coarse fabrics, including sack material, canvas, ropes, and twines.

### Jute

Jute fibre comes from a herbaceous annual plant which grows as high as 20 feet. The fibres are extracted from the plant stalk in a manner similar to flax. Jute tends to be brown in colour due to about 20 per cent lignin present in the fibre, but it does have a silklike luster. The individual cells of jute are very short, being about 0.1 inch long, and the fibre cross section has five or six sides. Jute is not as strong or as durable as flax or hemp. Jute fibres are stiff, but it does have an unusually high moisture regain. Jute's low cost, moderate strength, and availability make it an important fibre for use in sacks, bags, and packing materials. Fabrics made of jute are called burlap.

### Ramie

Ramie is a bast fibre often referred to as china grass. Ramie fibre is removed from the plan t by peeling or removing the bark and soaking the fibres in water, followed by scraping. The fibre must be further degummed by treatment in base before spinning. Ramie is a white fibre with excellent strength and luster. The fibre is stiff and fairly coarse. The cells of the fibre are very long, and the cross section irregular in shape. Ramie is useful in industrial applications and is being used in furnishings where rough, irregular fabrics are desired.

## STRUCTURAL PROPERTIES AND FORMATION OF RAYON

### RAYON

Rayon was the first man-made fibre to be produced commercially, and several types of rayon fibres are produced. Rayon is regenerated cellulosic material produced by solution of a cellulose source, followed by forcing of the solution through a spinneret and subsequent regeneration to form the fibre. The Federal Trade Commission defines rayon as a manufactured fibre composed of regenerated cellulose in which substituents have not replaced more than 15 per cent of the hydroxyl hydrogens. Rayon is produced by three methods to give the viscose rayons, cuprammonium rayon, and saponified cellulose acetate. Of these, viscose rayon is the most important and the most inexpensive to produce.

#### *Viscose Rayon*

Viscose rayon fibres are produced by the viscose process, which involves formation of soda cellulose, reaction with carbon disulfide to form cellulose xanthate, and spinning into dilute acid to regenerate the cellulose as rayon fibre. In practice the cellulose pulp is steeped in warm concentrated sodium hydroxide and pressed to remove excess sodium hydroxide. After aging, during which some degradation and breaking of cellulose chains occurs, the soda cellulose crumbs are mixed with carbon disulfide, and orange sodium cellulose xanthate is formed. The xanthate is then dissolved in dilute sodium hydroxide. At this point titanium dioxide is added to the solution if a delustered fibre is desired. The solution is aged and ripened until a proper viscosity is reached; the solution is then forced through the spinneret into dilute sulfuric acid to decompose the xanthate and regenerate the cellulose by the wet spinning process.

*In addition to acid, the following additives are often found in the spinning bath:*

- Sodium sulfate,
- Zinc sulfate, and
- Glucose.

Initially, the outer portion of the xanthate decomposes in the acid bath and a cellulose skin forms on the fibre. The sodium and zinc sulfates affect the rate of xanthate decomposition and fibre formation, while the action of zinc sulfate is confined to crosslinking the outermost portion of the fibre. The glucose tends to slow the rate of regeneration of the fibre and tends to make the fibre more pliable. The concentration and type of additives in the bath and the nature of the original cellulose will affect the physical properties of the rayon.

By careful control of the coagulation bath followed by mechanical stretching, crimp can be produced in the fibre. Slow regeneration and stretching of the rayon will tend to introduce greater areas of crystallinity within the fibre. High-tenacity rayons are formed in this manner. Also, the fibre can be modified through addition of other crosslinking agents to the spinning bath. High-wet-modulus rayons have been developed in recent years. These fibres are produced from high-grade cellulose starting materials, and the formation and decomposition of cellulose xanthate is carried out under the mildest of conditions to prevent degradation of the cellulose chains. After initial skin formation, the core of the fibre decomposes, hardens, and shrinks, causing wrinkling on the fibre skin. The final fibre cross section of viscose rayon appears serrated and irregular, and the nature of the crystallites in the skin and core will differ.

**Fig.** Viscose Rayon.

This skin effect can be controlled sufficiently to give nearly round fibre cross sections where carefully controlled coagulation occurs. The majority of viscose rayons have serrated and irregular cross sections due to the skin effect. The viscose rayon fibre is long and straight unless the fibre has been crimped, and striations due to the irregular cross section will run the length of the fibre. If the fibre has been delustered or spun dyed during fibre formation, particles of pigment will appear in the fibre.

Normal viscose fibres will generally consist of 25 per cent30 per cent crystalline areas within the fibre, and the average degree of polymerisation of glucose units in the cellulose chains will be 200-700. The crystallites in viscose rayon are somewhat smaller than those found in cotton. Hi gher-tena city rayons and high-wet-modulus rayons tend to be more crystalline, and higher degrees of polymerisation are found. The cross sections of these rayons tend to be more nearly round. The degree of crystallinity of high-wet-modulus rayons can reach 55 per cent.

***Cuprammoni urn Rayon***

Cuprammoni urn rayon is produced by solution of cellulosic material in cuprammonium hydroxide solution at low temperature under nitrogen, followed by extrusion of the solution through a spinneret into water and then sulfuric acid to decompose the cuprammonium complex and regenerate the cellulose. Cuprammonium rayon is more silklike than any of the other celluloses, but the cost of production is correspondingly higher. Cuprammonium rayon has a smooth surface, and no markings or striations are found on the fibre. The fibre cross section is nearly round.

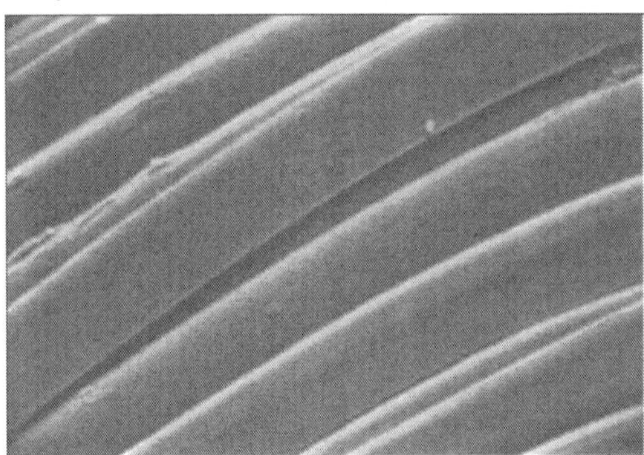

**Fig.** Cuprammonium Rayon.

***Saponified Cellulose Acetate***

Rayon can be produced from cellulose acetate yarns by hydrolysis in base, a process called saponification. Cellulose acetate molecules are unable to pack closely within the fibre due to lack of hydrogen bonding and the presence of bulky acetate side chains within the molecule. Since cellulose acetate is more plastic in nature, it can therefore be drawn and extended to give a high degree of orientation to the fibre. On hydrolysis the highly oriented cellulosic chains form hydrogen bonds with adjacent chains to give a rayon of high orientation and crystallinity. Saponified cellulose acetate has a nearly round but lobed cross section. The indentations caused by the lobes appear as light striations running the length of the filament.

**Physical Properties**

Since rayons are essentially pure cellulose, they would be expected to have physical properties similar to cellulosic fibres of natural origin. Differences in properties would be expected to depend on the degree of polymerisation, crystallinity, and orientation within the fibre. The dry and wet tenacities of the rayons vary over a wide range dependent on the degree of polymerisation and crystallinity with tenacities of 2 to 6 g/d dry and 1 to 4 g/d wet. For the more crystalline and oriented rayons, the drop in tenacity observed on wetting of the fibre is lower; however, none of the rayons exhibit higher wet strength than dry strength, as found with cotton. The great loss in strength of wet regular tenacity rayon makes it subject to damage during laundering.

The percentage elongation at break varies from 10 to 30 per cent dry and 15 to 40 per cent wet. The recovery at 2 per cent elongation ranges from 70 to 100 per cent. In general, the rayons have significantly higher elongations at break than observed for cotton. As the degree of crystallinity and orientation of the rayon increases, the elongation at break is lower. Rayons generally have somewhat better elastic properties at low elongations than is found for cotton. The rayons possess greater luster than cotton, and often delusterants are added to rayon prior to spinning to give a more subdued fibre. Since the rayons are less crystalline and oriented than cotton, they tend to swell more in water and undergo greater elongation under tension in both the wet and dry states.

During weaving and wet finishing considerable stresses can be introduced into many rayon fabrics, and relaxation of these stresses will be necessary before a dimensionally stable fabric is obtained. Owing in part to the more extensive networks of amorphous regions found in rayons, the moisture regain of rayon is significantly higher than that of cotton. The regains of rayons under standard conditions range from 11 per cent to 13 per cent. The lower crystallinity and degree of polymerisation of rayons also affect the way water acts on the fibres. Rayons as a consequence are swollen to varying degrees due to increased susceptibility to water penetration.

During wetting rayon may increase up to 5 per cent in length and swell up to double its volume. The heat conductivity and electrical properties are the same as those found for cotton. Rayon is cool to the touch, and static charges do not build up in the fibre at humidities greater than 30 per cent. Although rayon swells in water, it is not attacked by common organic solvents. It does dissolve in cuprammonium solutions.

The rayons are moderately stiff fibres with poor resiliency and wrinkle recovery properties. As in the case of cotton, resin treatments will effectively increase the resiliency of rayon. Often such treatments will tend to be more effective on rayon than on cotton due to the greater accessibility of the interior of rayon to the resin.

The specific gravities of viscose and cuprammonium rayons are the same as cotton and vary between 1.50 and 1.54. Only saponified cellulose acetate has a markedly different specific gravity.

**Chemical Properties**

The chemical properties of the rayons are essentially the same as those found for cotton. Cold dilute or concentrated base does not significantly attack rayon, but hot dilute acids will attack rayon at a rate faster than found for cotton. Although generally resistant to oxidising bleaches, rayon is significantly attacked by hydrogen peroxide in high concentrations.

Although silverfish are known to attack rayon, microorganisms which cause mildew do not readily attack rayon, except under more severe hot and moist conditions. Prolonged exposure of rayon to sunlight causes degradation of the cellulose chains and loss of strength of the fibres. The rayons are not thermoplastic, and they, like cotton, decompose below their melting point. Regular-tenacity rayon begins to decompose and lose strength at 150aC for prolonged periods and decomposes rapidly at 190°-210°C. High-tenacity rayons tend to decompose at slightly higher temperatures.

**End-Use Properties**

Rayon fibres found in consumer goods are known by numerous trade names. Regular- and high-tenacity viscose rayons are marketed as rayon or with names like Zantrel, Avril, Enkaire, and Fibro. Cuprammonium rayon and saponified cellulose acetate are not longer in production in the U.S. Regular- and medium-tenacity viscose rayons are among the least expensive of the man-made fibres, whereas high-strength, cuprammonium, and saponified cellulose acetate rayons are more expensive due to the greater care and additional steps necessary in manufacture of these fibres.

Although the properties of rayons are very nearly those of cotton, a greater range of properties exist within the various types of rayons available. Being an inexpensive fibre, rayon plays a role similar to that of cotton; however, rayon differs from cotton in that it can be modified to some degree during manufacture and is not as subject to world economic and climatic conditions. The dry strength of regula r- tenacityviscose and cuprammoni urn rayons are lower than that found for cotton, whereas high-tenacity viscose, polynosic, and saponified cellulose acetate rayons are significantly stronger than cotton. All rayons lose strength when wet and are more susceptible to damage while wet.

The abrasion resistance of rayon is fair, and abrasion of rayon fabrics becomes noticeable after repeated usage and launderings. The rayons are resistant to pill formation. Rayon fabrics wrinkle easily and without chemical treatment show poor crease recovery and crease retention. Durable press and wash-and-wear finishes tend to cause less degradation of rayon than found with cotton. Rayon possesses excellent moisture absorbency characteristics, but the weaker rayons swell and deform somewhat in the presence of moisture.

The hand of lower- strength viscose rayons is only fair, whereas the hand of cuprammonium and high-wet-modulus rayons is crisp and more desirable.

Fabrics of rayon are generally comfortable to wear and are of acceptable appearance. The rayons are resistant to common household solvents and to light and heat except under extreme conditions or prolonged exposures. Rayons can be dyed readily by a wide variety of dyes, and the colourfastness of dyed rayons is satisfactory. Rayons are moderately resistant to deterioration by repeated laundering, and a moderate degree of care is necessary for maintenance of rayon fabrics.

The lower-strength rayons exhibit poor dimensional stability during laundering. Rayons are affected significantly by detergents or other laundry additives. Greater care must be taken when ironing rayon fabrics than would be necessary with cotton. Viscose and cuprammonium rayons may be safely ironed at 176°. The rayons possess poor flammability characteristics and ignite readily on contact with a flame, as is the case for natural cellulosics. The versatility of rayons, coupled with their lower price, makes them suitable for many textile applications.

Rayon is used in clothing and home furnishings. Disposable non-woven garments and products of rayon have been introduced to the consumer in recent years. The stronger rayons have been used in tire cord for several decades but have lost a significant portion of this important market in recent years. Rayon ha s been used more and more in blends with synthetic fibres, since rayon undergoes less degradation than cotton with durable press and wash-and-wear finishes.

## ACETATE AND TRIACETATE IN CELLULOSE ESTER FIBRES

The cellulose esters triacetate and acetate are the two major fibres of this type. The production of acetate fibres resulted from an attempt to find a new outlet for cellulose acetate used as aircraft "dope" in World War 1. By 1921, the first acetate fibres were being produced. Although small quantities of cellulose triacetate fibres were produced before World War I, it was not until after 1954 that cellulose triacetate fibres were produced commercially in large quantities.

The Federal Trade Commission defines the acetate fibres as manufactured fibres in which the fibre-forming substance is cellulose acetate. Triacetate fibres are those cellulose fibres in which more than 92 per cent of the hydroxyl groups are acetylated. The term secondary acetate has been used in the past in reference to the acetates being less than 92 per cent acetylated. In recent years, these fibres have simply gone by the designation acetate. Although triacetate and acetate are very similar structurally, this slight difference in degree of esterification provides significant differences in properties for these fibres.

### Structural Properties and Formation

The cellulose esters triacetate and acetate are formed through acetylation of cotton linters or wood pulp using acetic anhydride and an acid catalyst in acetic acid. At this point the cellulose is fully acetylated. The triacetate is precipitated from solution by addition of water and dried. The cellulose triacetate is normally dissolved in methylene chloride to form about a 20 per cent solution and forced through the spinneret into a stream of hot air to evaporate the solvent to form triacetate fibres by dry spinning. Acetate is formed by allowing the water-precipitated cellulose triacetate solution above to stand as long as 20 hours in aqueous solution.

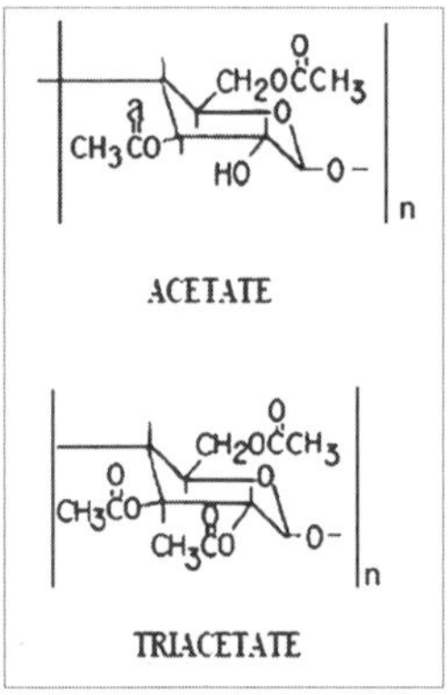

ACETATE

TRIACETATE

During this ripening process, some of the acetyl groups are hydrolysed or removed from the cellulose. The acetate is then washed, dried, and dissolved in acetone to form the spinning "dope." The "dope" is forced through a spinneret into hot air to form the acetate fibre by dry spinning. Triacetate and acetate fibres can also be formed by wet spinning processes. In triacetate, essentially all hydroxyl groups of the cellulose have been acetylated, whereas 2.3-2.5 hydroxyl groups per anhydroglucose unit have been acetylated in cellulose acetate. In both triacetate and acetate, hydrogen bonding between cellulose chains for the most part is eliminated. Also, the bulky acetate group discourages close tight packing of adjacent cellulosic chains.

As would be expected, these changes in molecular structure greatly affect the physical properties of triacetate and acetate. Although limited hydrogen bonding occurs between molecules within acetates, triacetate is incapable of forming hydrogen bonds. Van der Waals forces caused by the interaction of adjacent acetate and triacetate chains are the major associative forces between the acetylated cellulose molecular chains.

The average number of acetylated anhydroglucose units in chains of acetates and triacetates usually varies between 250 and 300 units. The microscopic structures of triacetate and acetate are similar. The fibre cross section of each fibre is irregular, with as many as five or six lobes. The longitudinal views of these fibres show striations running the length of the fibre. Small particles of delusterant will be visible in these fibres if they have been added to the fibre prior to spinning. Acetate and triacetate fibres are very similar in appearance to the regular-tenacity viscose rayons.

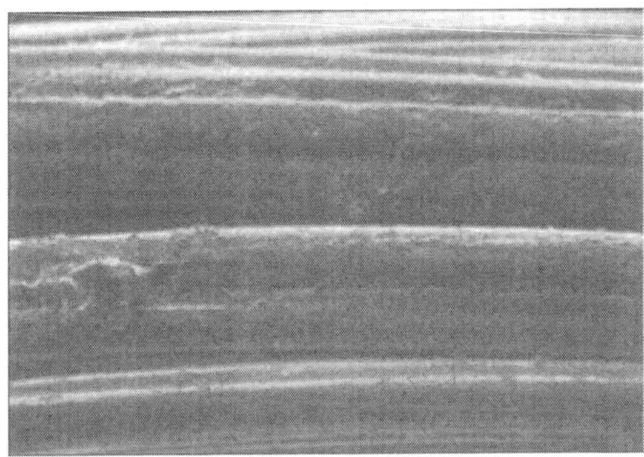

**Fig.** Acetate.

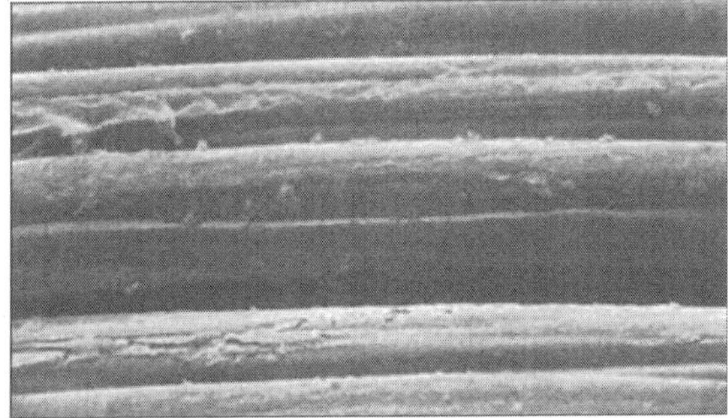

**Fig.** Triacetate.

**Physical Properties**

Since acetate and triacetate have only 1imited associative forces between molecular chains, they are considerably weaker than most of the cellulosic fibres. The tenacity of acetate and triacetate falls in the range 1-1.5 g/d. On wetting, their tenacities fall to 0.8-1.2 g/d. The elongations at break for acetate and triacetate are 25 per cent-40 per cent. On wetting, acetate breaks at 35 per cent-50 per cent elongation, whereas triacetate breaks at 30 per cent-40 per centelongation.

These fibres possess good recoveries at low elongations. Above 5 per cent elongation, plastic deformation of the fibres occurs and subsequent recovery is poor. Acetates and triacetates are moderately stiff fibres and possess good resiliency on bending and deformation, particularly after heat treatment. The specific gravities of acetate and triacetate range from 1.30 to 1.35. The moisture regains of these fibres are markedly lower than found for rayon and cotton. Acetate and triacetate have moisture regains of 6.0 per cent and 4.5 per cent, respectively. On heating, the moisture regain of triacetate drops further to 2.5 per cent. Acctate and triacetate fibres conduct heat readily and are cool to the touch. Being thermoplastic, acetate and triacetate fibres soften at 200°C and 225°C and melt at 232°C and 300°C, respectively. Triacetate can be heat set more readily than acetate, and permanent pleats can be satisfactorily heat set into triacetate fabric. Heat treatment of triacetate will change the character of the fibre and give a more dense molecular structure.

Acetate and triacetate possess excellent electrical characteristics and can be used as insulating materials. As a consequence, static buildup is a problem on these fibres unless antistatic treatments are applied. Acetates and triacetates are attacked by a number of organic solvents capable of dissolving esters. Acetate swells or dissolves in acetone, ethyl acetate, phenol, chloroform, and methylene chloride, whereas triacetate dissolves in formic acid, acetic acid, methylene chloride, or chloroform.

**Chemical Properties**

Acetate and triacetate are resistant to dilute acids but are attacked by concentrated acids causing hydrolysis and removal of the acetate ester groups. While resistant to dilute alkalies, acetate is more susceptible to attack by alkalies than triacetate. Dilute alkali solutions attack acetate causing saponification and removal of the acetate groups on the fibres. Acetate is attacked by strong oxidising agents, while triacetate has somewhat better resistance to attack by these reagents. Both fibres, however, can be bleached under proper conditions. Both acetate and triacetate are not attacked by insects or microorganisms. Acetate and triacetate are resistant to attack by sunlight, with greater resistance found for triacetate. Acetate and triacetate will withstand long periods at elevated temperatures below their melting points without significant loss of strength.

**End-Use Properties**

Acetates and triacetates are relatively inexpensive and have certain properties which are desirable in selected end-use applications. Acetate is marketed under trade names such as Airloft, Estrom, and Loftura, while triacetate fibres bear the trade names Tricel and Arnel. Both acetate and triacetate are weak fibres having low tenacities, and they cannot be used in applications where high strength is required unless they are blended with other fibres.

The high elongations at break found for acetate and triacetate help compensate for these fibres' low strengths to some extent, and the fibres exhibit good recoveries at low extensions. The abrasion resistance of acetate and triacetate is poor, and these fibres cannot be used in applications requiring high resistance to rubbing and abrasion; however, the resistance of these fibres to pilling is excellent. While acetate and triacetate are moderately absorbent, their absorbencies cannot compare with the pure cellulosic fibres. The hand of acetate fabrics is somewhat softer and more pliable than triacetate, which possesses a crisp firm hand. Fabrics of both fibres possess excellent draping characteristics.

Fabrics of acetate and triacetate have a pleasing appearance and a high degree of luster, but the luster of these fabrics can be modified through addition of delusterants. Both acetate and triacetate are susceptible to attack by a number of household chemicals. Acetate and triacetate are attacked by strong acids and bases and by oxidising bleaches. Acetate has only fair sunlight resistance, whereas the sunlight resistance of triacetate is superior. Both fibres have good heat resistance below their melting points. Acetate and triacetate cannot be dyed by dyes used for cellulosic fibres.

These fibres can be satisfactorily dyed with disperse dyes at moderate to high temperatures to give even, bright shades. The laundry colourfastness of triacetate is excellent; however, dyed acetate fabrics generally have moderate to poor colourfastness. The acetate fibres generally require a certain degree of care in laundering, although they exhibit greater wrinkle recovery on drying than cellulosic fibres. Both acetate and triacetate fabrics do not shrink or lose their shape readily during laundering and can be laundered satisfactorily under mild laundry conditions. Both acetate and triacetate may be satisfactorily dry cleaned in Stoddard solvents, but certain other solvents must be avoided due to fibre solubility. Acetate and triacetate dry quickly and may be tumble dried or drip dried. During tumble drying of these materials it is essential that the dryer be cool during the final tumbling. Both acetate and triacetate can be safely ironed, but direct contact of the fabric with the iron is not recommended. Acetate must be ironed at a lower temperature than triacetate, and the maximum safe ironing temperature for these fabrics is 175 °C and 200°C, respectively.

The flammability characteristics of these fibres are similar to those of the cellulosic fibres, and they have LOIs of 18. Acetate is used in dresses, blouses, foundation garments, lingerie, garment linings, some household furnishings, and certain specialty fabrics. Triacetate is used in sportswear, tricot fabrics, and in garments where pleats and pleat retention is important, as well as in certain specialty fabrics.

## VINYL FIBRES

*Vinyl fibres are those man-made fibres spun from polymers or copolymers of substituted vinyl monomers and include:*

- Vinyon,
- Vinal,
- Vinyon-vinal matri x,
- Saran, and
- Polytetrafluoroethylene fibres.

Acrylic, modacrylic and polyolefin are also formed from vinyl monomers, but because of their wide usage and particular properties they are usually considered as separate classes of fibres. Thc vinyl fibres are generally specialty fibres due to their unique properties and uses. All of these fibres have a polyethylene hydrocarbon backbone with substituted functional groups that determine the basic physical and chemical properties of the fibre.

## VINYON

Vinyon is defined as a fibre in which at least 85 per cent of the polymerised monomer units are vinyl chloride. Vinyon fibres have high chemical and water resistance, do not burn, but do melt at relatively low temperatures and dissolve readily in many organic solvents, thereby limiting their application.

### Structural Properties

Vinyon is formed by emulsion polymerisation of vinyl chloride or by copolymerisation of vinyl chloride with less than 15 per centvinyl acetate in the presence of free radical catalysts to a DP of 200-450:

$$\left[ (CH_2\underset{Cl}{\underset{|}{C}}H)_x(CH_2\underset{OCCH_3}{\underset{|}{\overset{O}{C}}}H)_y \right]_n$$

$$x > 85\% ,\ y < 15\%$$

VINYON

The polymer is precipitated and isolated by spray drying and then dissolved in acetone; the polymer solution is dry spun to form the fibre. Comonomers such as vinyon cetate are added to reduce the crystal linity of the drawn fibre and to increase the amorphous areas within the fibre. Weak hydrogen bonding between chlorine and hydrogen on adjacent vinyon chains would be expected with tight packing of the molecular chains in the absence of comonomer. The fibres are spun in nearly round or dog-bone cross section.

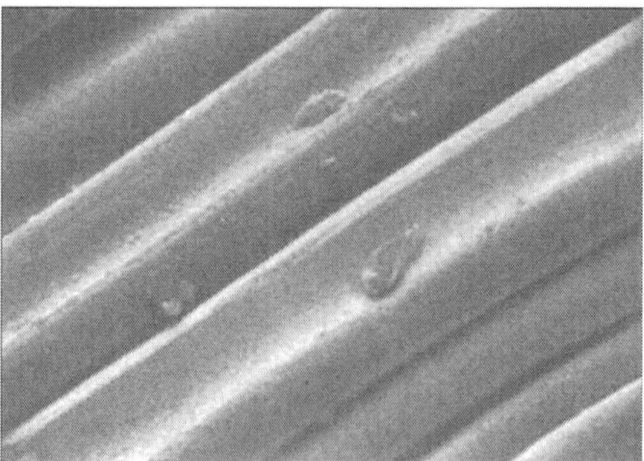

**Fig.** 10-1. Vinyon.

### Physical Properties

Vinyon fibres have a strength of 1-3 g/d in both the wet and dry states; elongations at break vary between 10 per cent and 125 per cent. At low elongations vinyon fibres recover completely from deformation. Vinyon fibres are soft and exhibit good recovery from bending deformation. The fibre has moderate density, with a specific gravity of 1.33-1.40. Vinyon is extremely hydrophobic, having a moisture regain of 0.0 per cent-0.1 per cent under standard conditions. Vinyon is readily dissolved in ketone and chlorinated hydrocarbon solvents and is swollen by aromatic solvents. The fibre is a poor heat and electrical conductor and possesses potential in insulation applications. Unfortunately, the oriented fibre softens and shrinks at temperatures above 60 ° C, thereby limiting the applications in which it may be used.

**Chemical Properties**

Vinyon fibre is chemically inert and possesses chemical properties similar to polyolefin fibre. Vinyon is only slowly attacked by ultraviolet rays in sunlight. Vinyon fibre melts with decomposition at 135°-180°C, with vinyon containing comonomer having a lower melting/decomposition temperature.

**End-Use Properties**

Vinyon as pure polyvinyl is marketed as PVC-Rhovyl, while vinyon HH is a copolymer. The fibre is of low strength but has properties that make it useful in apparel where heat is not a factor. It is difficult to dye. It may be laundered readily, but it is attacked by convnon dry-cleaning solvents. Since the fibre may not be heated above 60°C, it may be tumble dried only at the lowest heat settings.

The fibre is non-flammable and does not support flaming combustion. Vinyon finds its major use in industrial fabrics including filters, tarps, and awnings, in protective clothing, and in upholstery for outdoor furnishings. It is also used as a bonding fibre in heat bonded non-wovens and as tire cord in specialty tires.

**VINAL**

Vinal fibres are made from polymers containing at least 50 per cent vinyl alcohol units and in which at least 85 per cent of the units are combined vinyl alcohol and acetal crosslink units. The fibre is inexpensive, resembles cotton in properties, and is produced in Japan.

**Structural Properties**

Vinal fibres are formed from inexpensive starting materials through a complex process. Vinyl acetate monomer is solution polymerised to polyvinyl acetate using free radical initiation techniques.

The polyvinyl acetate is hydrolysed to polyvinyl alcohol in methanol under basic conditions. Aqueous polyvinyl alcohol solution is wet or dry spun followed by exposure to an aldehyde such as formaldehyde or benzaldehyde and heat to form acetal crosslinks between chains to insolubilise the fibre. The polyvinyl alcohol chains are tightly packed except near the juncture of the acetal crosslinks. Extensive hydrogen bonding between hydroxyl groups on adjacent chains occurs, which contributes to its highly packed structure, and the fibre is about 50 per cent crystalline:

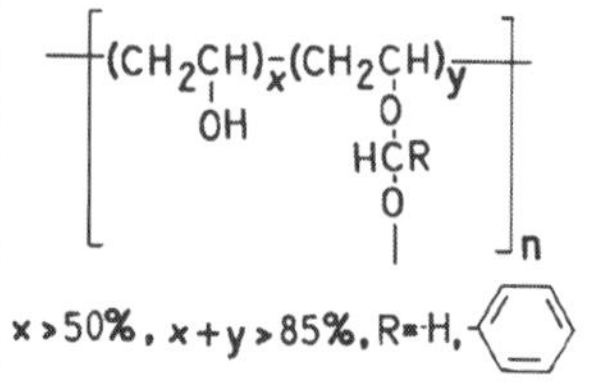

$$\left[ (CH_2\underset{\displaystyle OH}{CH})_x (CH_2\underset{\displaystyle \overset{|}{O}-\overset{|}{HCR}-\overset{|}{O}}{CH})_y \right]_n$$

$x > 50\%$, $x + y > 85\%$, R = H, $-C_6H_5$

VINAL

The fibre surface is somewhat rough with lengthwise striations and possible periodic twists. The fibre may be nearly round or U-shaped in cross section with a noticeable skin and core.

**Physical Properties**

Vinal fibre is moderately strong with a dry tenacity of 3-8.5 g/d and possesses a moderate extension at break of 10 per cent-30 per cent. On wetting the strength of the fibre decreases and the elongation at break increases slightly. The fibre exhibits poor to moderate recovery from elongations as low as 2 per cent. The fibre is moderately stiff, but does not recover readily from dry or wet deformation.

The fibre has a specific gravity of 1.26-1.30. Vinal fibre has a moisture regain of 3.0 per cent-5.0 per cent under standard conditions, and is swollen and attacked by aqueous phenol and formic acid solutions. The fibre is a moderate heat and electrical conductor. On heating to 220°-230° the fibre undergoes 10 per cent or more heatinduced shrinkage.

**Chemical Properties**

Vinal is somewhat sensitive to acids and alkalies. It is attacked by hot dilute acids or concentrated cold acid solutions with fibre shrinkage and is yellowed by strong aqueous solutions of alkali. Oxidising, reducing, and biological agents have little effect on vinal. Sunlight causes vinal to slowly lose its strength with perceptible changes in colour. At 230 °-250° vinal shrinks and softens with decomposition.

**End-Use Properties**

Vinal resembles cotton and other cellulosics in end-use properties. Kuralon and Manryo are names under which vinal fibres are marketed. Vinal fibre has good strength and excellent abrasion and pilling resistance. Like cellulosics, vinal breaks at low elongations, exhibits poor recovery from small deformations, and wrinkles readily unless treated with durable press resins of the type used for cellulosics.

Fabrics of vinal have a warm comfortable hand, are absorbent, and exhibit good drapabi1ity. The fibre has a silklike appearance and luster. It has excellent sunlight resistance and fair heat resistance. It dyes readily with dyes for cellulosics. Its launderability and dry-cleanability are very good, and it dries more readily than cellulosics. Vinal may be ironed dry up to 120°C, but undergoes severe deformation if ironed wet below 120°C. Vinal burns readily like a cellulosic fibre and has a LOI of 20. Vinal is used in apparel as well as industrial applications where an inexpensive fibre of moderate strength and properties is desired. It is useful in fibre blends to increase comfort and aesthetic properties.

## VINYON-VINAL MATRIX FIBRE

In response to the need for a fibre of low flammability and low toxic gas formation on burning, Kohjin company developed and marketed a vinyon-vinyl matrix fibre under the trade names Cordela and Cordelan. The fibre is believed to be formed through grafting of vinyl chloride to polyvinyl alcohol followed by mixing of the resultant copolymer with additional polyvinyl alcohol. The polymer mixture is wet spun, oriented, and crosslinked with aldehydes.

The fibre has a kidney-shaped cross section, and no outer skin is evident. The fibre is moderately strong with a dry tenacity of about 3 g/d and wet tenacity of 2.2 g/d. The elongation at break is 15 per cent-20 per centwet or dry, with poor recovery from even low elongations. It has a specific gravity of 1.32 and moisture regain of 3.0 per cent.

Besides its low flammability, it has good chemical resistance except under extreme conditions. The fibre is heat sensitive above 100 °C, but it may be dyed at lower temperatures with a large range of dyes for cellulosics as well as man-made fibres. Vinyonvinyl matrix fibre is used primarily in applications where low flammability in apparel is desired and has been used extensively in children's sleepwear.

## SARAN

Saran is the generic name for fibres made from synthetic copolymers that are greater than 80 per cent vinylidene chloride. The fibre is formed through emulsion copolymerisation of vinylidene chloride with lesser amounts of vinyl chloride using a free radical catalyst, and the precipitated copolymer is melt spun into the fibre:

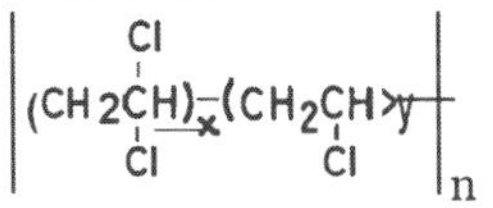

$$\left[ (CH_2CCl_2)_x (CH_2CHCl)_y \right]_n$$

$x > 80\% \cdot y < 20\%$

SARAN

The fibre is of moderate crystallinity and resembles vinyon in properties. The fibre has wet and dry strength of 1-3 g/d, elongation at break of 15 per cent-30 per cent, and good recovery from less than 10 per cent elongation. The fibre has a specific gravity of 1.7 and has excellent resiliency. It has essentially no affinity for water, but it is soluble or attacked by cyclic ethers, ketones, and aromatic solvents. It is a good heat and electrical in sulator, but it softens at 115°-150°C and melts at 170°C.

The fibre is chemically inert and only slowly affected by the ultraviolet rays in sunlight. The fibre is of low flammability. Saran is marketed as saran or under the trade name Rovana. The fibre has high resiliency and low water absorbency and is chemically inert and used in applications where these qualities are desirable and where heat sensitivity is not a problem. The fibre is generally not used in apparel but rather in automobile upholstery, outdoor fabrics, home furnishings, and industrial applications.

## POLYTETRAFLUOROETHYLENE

Tetrafluoroethylene is better known by its trade name—Teflon—and is widely used in many applications including specialty fibres. Poly tetrafluoroethylene fibre is extremely hydrophobic and chemically and thermally stable and is used in applications where such stability and inertness is needed. Polytetrafluoroethylene fibres are produced through polymerisation of tetrafluoroethylene under conditions of high temperature and pressure using peroxide catalysts. The polymer is spun as a dispersion to form a weak fibre which is then heated at 385°C to fuse the individual fibre particles.

The fibre must be bleached to give a white fibre. The highly electronegative fluorine atoms in the fibre result in an extremely tight packing of adjacent molecular chains tightly held by van der Waals forces and with a high degree of crystallinity. The fibres are smooth with a round cross section. The fibre has a tenacity of 1-2 g/d and elongations at break vary from 15 per cent to 30 per cent. It is a soft flexible fibre of high density, and it is totally hydrophobic. It is unaffected by solvents except for perfluorinated hydrocarbons above 300°C.

It is an excellent electrical and heat insulator and is not affected by heat up to 300°C. It is chemically inert and may be used in numerous industrial end-use applications including protective clothing. Its hydrophobicity has permitted is to be used to form breathable but water repellant composite materials for textile usage, particularly in outdoor and rainwear.

### Elastomeric Fibres

Elastomeric fibres are those fibres that possess extremely high elongations at break and that recover fully and rapidly from high elongations up to their breaking point. Elastomeric fibres are made up of molecular chain networks that contain highly amorphous areas joined together by crossl inks. On elongation, these amorphous areas become more oriented and more crystalline in nature.

Elongation continues until the crosslinks in the structure limit further molecular movement. If additional force is placed on the elastomeric fibres at this point, molecular scission occurs, causing a reduction in properties and ultimate breaking of the fibre.

*Elastomeric fibres include:*

- The crosslinked natural and synthetic rubbers,
- Spandex fibres,
- Anidex fibres and the side-by-side biconstituent fibre of nylon and spandex.

The fibres are all used in specialised applications where high elasticity is necessary within the textile structure.

## RUBBER

Rubber fibres from natural sources have been known for over 100 years. Natural rubber in commerce is derived from coagulation of Hevea brasiliensis latex and is primarily polyisoprene, a diene polymer. Most synthetic rubbers were developed during and following World War II. They are crosslinked diene polymers, copolymers containing dienes, or amorphous polyolefins. Both the natural and synthetic rubbers must be crosslinked with sulfur or other agents before true elastomeric properties are introduced. In addition, accelerators, antioxidants, fillers, and other materials are added to the polymeric rubber prior to fibre formation. Rubber fibres exhibit excellent elastic properties but are sensitive to chemical attack, thereby limiting their usefulness.

### Structural Properties

*Rubber fibres are derived from the sources outlined above and structurally are:*

- Crosslinked polyisoprene,
- Polybutadiene,
- Diene-monomer copolymers, or
- Amorphous polyolefins.

Less common or special use rubbers include the acrylonitrile-diene copolymer Lastrile and the chloroprene polymer Neo-prene. Structural formuli of typical rubbers follow:

The rubber polymers emulsified as a latex are blended with vulcanising agents, chemical accelerators, antioxidants, fillers, and other materials. The polymer blend is extruded or coagulated and cut into fibres and heated to crosslink the structure. Of necessity, the rubber fibres are large in cross section and have a higher linear density compared to other man-made fibres. The hydrocarbon chains in the crosslinked rubbers are in a highly folded but random and amorphous configuration, but they are attached periodically to adjacent polymer chains through multiple sulfur or other crosslinks.

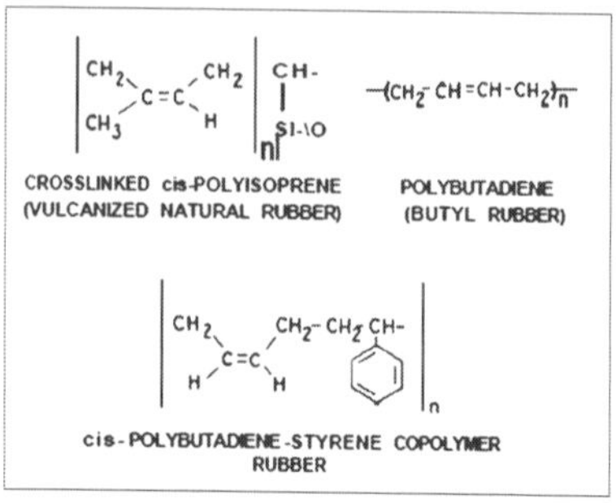

CROSSLINKED cis-POLYISOPRENE (VULCANIZED NATURAL RUBBER)

POLYBUTADIENE (BUTYL RUBBER)

cis-POLYBUTADIENE-STYRENE COPOLYMER RUBBER

On stretching, the molecular chains in the amorphous region untangle and straighten and orient into parallel, more crystal line structures, up to the elastic limit determined by the crosslinks present. On relaxation, the molecular chains return to their lower-energy amorphous state. Rubber fibres tend to possess a square or round cross section depending on whether the fibres were extruded or cut.

## Physical Properties

Rubber fibres possess tenacities of only 0.5-1 g/d wet or dry. Elongation at break for these elastomeric fibres varies between 700 per cent and 900 per cent, with nearly complete elastic recoveries even at higher elongations. Because of their size, rubber fibres are moderately stiff but resilient. They are light, having a specific gravity of only 0.95-1.1. Certain environmental properties of rubber limit its usefulness. Being hydrocarbon material, rubber fibres do not absorb moisture and have a regain of 0.0 per cent under standard conditions. The fibre is a poor heat conductor and a good heat and electrical insulator. The fibre is readily swollen by ketones, alcohol, hydrocarbons, and oils and softens if heated above 100°C.

## Chemical Properties

Rubber fibres are quite susceptible to attack by chemical s both in solutions and in the environment due to the presence of extensive unsaturation in the molecule. Rubber is readily attacked by concentrated sulfuric and nitric acids and by oxidising agents such as hydrogen peroxide and sodium hypochlorite at elevated temperatures. Al though biological agents don't readily attack rubber, heating above 100°C or exposure to sunlight accelerates its oxygen-induced decomposition. Rubber fibres are particularly susceptible to attack by atmospheric ozone, a major component of smog.

## End-Use Properties

*Rubber fibres are manufactured by several companies as rubber fibre or under trade names such as:*

- Buthane,
- Contro,

- Hi-Flex,
- Lactron,
- Lastex, and
- Laton.

Rubber fibres show good elastomeric properties and reasonable aesthetic properties particularly as the core of a textile yarn structure. Rubber fibres usually have a dull luster due to the additives and fillers within the fibre. The fibre has poor resistance to household chemicals, sunlight, heat, and atmospheric contaminants. It cannot be dyed readily, and the polymer must be coloured prior to fibre extrusion and curing.

The fibre possesses only fair launderability and poor dry-cleanability. Owing to its low moisture absorbance, it dries readily, but care must be taken to dry or iron the fibre below 100°C. Rubber fibres burn readily but tend to be self-extinguishing. Rubber fibres are used extensively in applications where elastomeric fibres are desired, such as in swimwear, foundation garments, outerwear, and underwear, hosiery, and surgical bandages. The use of rubber fibres has decreased in recent years as more stable elastomeric fibres such as spandex have been introduced.

## SPANDEX

Spandex fibres are elastomeric fibres that are >85 per cent segmented polyurethane formed through reaction of a diisocyanate with polyethers or polyesters and subsequent crosslinking of polyurethane units. The spandex fibres resemble rubber in both stretch and recovery properties, but are far superior to rubber in their resistance to sunlight, heat, abrasion, oxidation, oils, and chemicals. They find the widest use of any of the elastomeric fibres.

### Structural Properties

Spandex is a complex segmented block polymer requiring a complex series of reactions for formation. Initially, low molecular weight polyethers and polyesters containing reactive terminal hydroxyl and/or carboxyl groups are reacted with diisocyanates by step growth polymerisation to form a capped prepolymer.

This polymer is melt spun or solvent spun from N,N-dimethyl-formamide into a fibre; then the fibre is passed through a cosolvent containing a reactive solvent such as water that reacts with the terminal isocyanite groups to form urethane crosslinks. Typical structures for spandex with urethane crosslinks follow:

The polyether or polyester segments in spandex are amorphous and in a state of random disorder, while urethane groups segmenting the polyether or polyester segments can form hydrogen bonds and undergo van der Waals interactions with urethane groups on adjacent chains. Chain ends will be crosslinked or joined to other chains through urea groups. On stretching, the amorphous segments of the molecular chains become more ordered up to the limit set by the urea linkages.

$$-\!\left[\mathrm{NH\overset{O}{\overset{\|}{C}}NH{-}R'{-}NH\overset{O}{\overset{\|}{C}}O{-}R{-}\overset{O}{\overset{\|}{C}}NH{-}R'}\right]_{\overline{n}}$$

$$R = \left(CH_2CH_2O\right)_{\overline{n}},\ (CH_2)_6\overset{O}{\overset{\|}{C}}O,\ \text{etc.}$$

R' = [aromatic ring structures, incl. $-C_6H_4-CH_2-C_6H_4-$ and methyl-substituted ring $CH_3$] etc.

**SPANDEX**

On relaxation, the fibre returns to its original, less ordered state. Spandex fibres are long, opaque fibres available in many cross sections including round, lobed, or irregular.

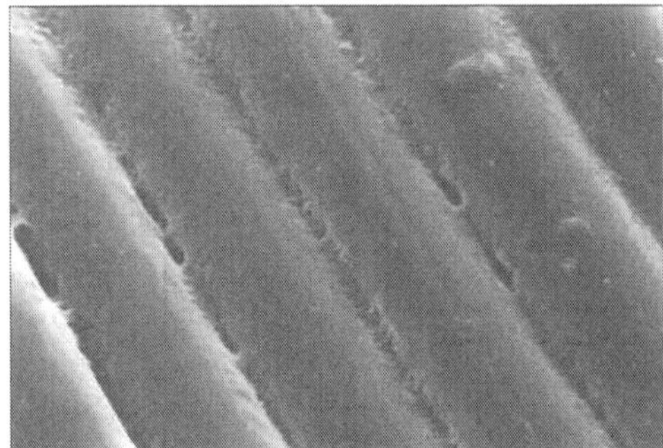

**Fig.** Spandex.

## Physical Properties

Spandex fibres are weak but highly extensible fibres with tenacities of 0.5-2 g/ d dry with slightly weaker wet tenacities, and elongations at break of 400 per cent-700 per cent. Spandex exhibits nearly complete recovery from high elongations and is quite resilient. The fibres have specific gravities of 1.2-1.4 and are somewhat more dense than rubber fibres. Owing to the more hydrophilic nature of spandex, its moisture regain varies between 0.3 per cent and 1.3 per cent under standard conditions. Although spandex is swollen slowly by aromatic solvents, it is generally unaffected by other solvent s. Spandex is a heat insulating material and shows poor heat conductivity. It softens between 150°C and 200 °C and melts between 230°C and 290° C. It has moderate electrical resistivity and builds up some static charge under dry conditions.

## Chemical Properties

Spandex fibres are much more stable to chemical attack than rubber. Spandex is attacked by acids and bases only at higher concentrations and under more extreme reaction conditions. Reducing agents do not attack spandex, but chlorine bleaches such as sodium hypochlorite slowly attack and yellow the fibre. Spandex generally resists attack by organic solvents and oils and is unaffected by biological agents. Sunlight causes slow yellowing and deterioration of spandex, and heat causes slow deterioration at elevated temperatures.

### End-Use Properties

*Spandex is widely used and sold as an elastomeric fibre under that designation or under such trade names as:*

- Cleerspan,
- Glospan,
- Lycra, or
- Numa.

Spandex possesses excellent elastomeric properties and acceptable aesthetics for use in constructions requiring such a fibre. Spandex is dull, but luster may be improved with brighteners. Although it has low water absorbency, spandex has good resistance to abrasion and wrinkling and may be heat set if desired.

It has good resistance to household chemicals other than bleach and is resistant to sunlight or heat-induced oxidation. It can be dyed relatively easily to reasonable degrees of colourfastness. It stands up well to repeated laundering and dry-cleaning, and it dries readily without damage at moderate temperatures. Spandex burns readily but melts and shrinks away from the flame. Spandex can be used effectively in all elastomeric fibre applications.

## OTHER ELASTOMERIC FIBRES

### Anidex

In 1970, anidex fibres were introduced as an elastomeric fibre by Rohm and Haas with the trade name Anim. Anidex fibres are defined as fibres containing polymers that are at least 50 per centof one or more polymerised acrylate esters. Anidex fibres are formed through emulsion copolymerisation of acrylate esters with reactive crosslinkable comonomers. The resulting copolymer ernulsion is mixed with a filler and wet spun to form a fibre which is heated to crosslink the polymer chains and provide the necessary elastomeric properties.

The morphology and elastomeric action of the fibre resemble spandex and rubber, but anidex generally has a lower elongation at break than these fibres. It has a round cross section. The fibre has a specific gravity of 1.22 and a moisture regain of 0.5 per cent under standard conditions. An index is reported to be more resistant to heat, light, and chemicals than either spandex or rubber. Otherwise, the fibre possesses end-use properties very much like those of spandex.

Anidex fibres apparently did not have sufficient differences in properties to become an economic success and are no longer being produced.

### Nylon-Spandex Biconstituent Fibre

An inherent problem with elastomeric fibres is their low strength and limited abrasion resistance. A new biconstituent elastomeric fibre-Monvelle— has been introduced to answer these problems. The fibre is a side-by-side bicomponent fibre containing nylon and spandex. The fibre is melt spun by special spinnerets to form the fibre. On drawing, the nylon portion of the fibre stretches and becomes more molecularly oriented, whereas the spandex simply elongates as would be expected for an elastomeric fibre. On relaxation, the spandex portion contracts, whereas the nylon portion does not, and the fibre coils into a tight coil which will have excellent elastomeric properties and high strength even as the spandex approaches its elastic limit. Sheer fibres may be formed from this bicomponent fibre, and the fibre is readily dyeable due to its nylon content. The fibre is particularly useful in support hose and in pantyhose and other constructions where elasticity and strength are important.

## MINERAL AND METALLIC FIBRES

A number of fibres exist that are derived from natural mineral sources or are manufactured from inorganic and mineral salts. These fibres are predominantly derivatives of silica or other metal oxides. In addition, metal fibres are produced. The common feature of these fibres is their inorganic or metallic composition and tendency to be heat resistant and non-flammable, with the exception of polymer-coated metallic fibres.

### GLASS

Fibres spun from glass are completely inorganic in nature and possess unique properties that cannot be found in organic textile fibres. Glass fibres have some deficiencies in properties that severely limit their use in apparel. Glass fibres are used in a number of industrial and aerospace applications and in selected home furnishing uses where heat and environmental stability are of prime importance.

### Structural Properties

Glass fibres are formed from complex mixtures of silicates and borosilicates in the form of mixed sodium, potassium, calcium, magnesium, aluminum, and other salts.

*The silicate-borosilicate prepolymer is prepared through mixing and fusing the following inorganic salts in the range of concentrations indicated:*

- *Silica:* 50 per cent-65 per cent
- *Calcium oxide:* 15 per cent-25 per cent
- *Alumina:* 2 per cent-18 per cent
- *Boron oxide:* 2 per cent-15 per cent
- *Other oxides:* 1 per cent-10 per cent

The exact composition of the manufactured glass polymer will affect the ultimate properties of the glass fibre formed. The glass thus formed as small marbles is melted in an electric furnace at high temperatures and melt spun to form fine smooth glass fibre filaments.

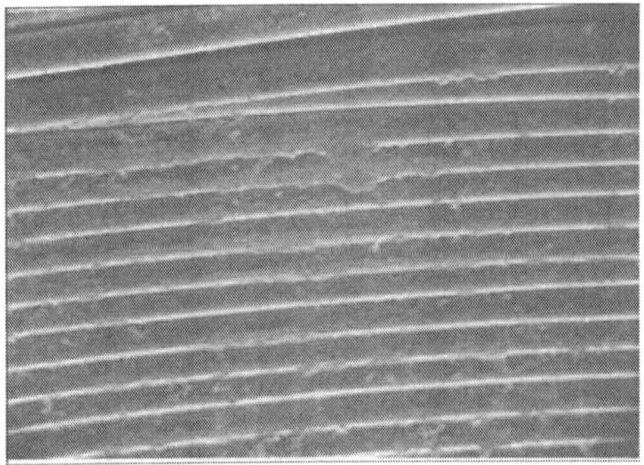

**Fig.** Glass.

The resultant silicate-borosilicate chains within the fibre are not ordered, and the glass structure is totally amorphous. Glass is actually a supercooled liquid that exhibits extremely slow but observable flow characteristics with time. The polymer chains within a glass fibre can be represented as follows:

$O^-$
O   $O^-Na^+$   $O^-K^+$   $O^-Na^+$   $O^-Mg^{++}$
—[ Si–O–Si–O–Si–O–Si–O–Si–O— ]$_n$
$O^-$   $Ca^{++}$ $O^-$   $O^-Na^+$   O   O
B

GLASS

## Physical Properties

Glass fibres are extremely strong, with tenacities of 6-10 g/d dry and 5-8 g/d wet. They possess elongations at break of only 3 per cent or 4 per cent but are perfectly elastic within this narrow deformation range up to their breaking point. Glass fibres are quite stiff and brittle and break readily on bending. As a result, they exhibit poor resistance to abrasion, although appropriate organic sizings can alleviate this problem to some extent. Glass fibres are extremely dense, having a specific gravity of about 2.5. The surface of glass fibres can be wetted by water, but otherwise the fibres have essentially no affinity for water. As a result the moisture regain of these fibres is 0.5 per cent or less. Glass fibres are not soluble in common organic solvents but can be slowly dissolved by concentrated aqueous base and more rapidly by hydrofluoric acid. Glass fibres exhibit excellent heat and electrical insulation properties and are not affected by heat up to their melting point of 750°C or above.

### Chemical Properties

Glass fibres are chemically inert under all but the most extreme conditions. Glass fibres are attacked and etched by dilute hydrofluoric acid and by concentrated alkalies over long periods of exposure. Oxidising and reducing agents and biological agents have no effect on glass fibres under normal conditions, and glass fibres are unaffected by sunlight or heat.

### End-Use Properties

Glass fibres are manufactured for industrial and consumer use under a number of names including fibreglas, Beta glass, J-M fibreglass, PPG fibreglass, and Vitron. Glass fibres are strong, but they exhibit poor abrasion resistance, which limits their use in textile structures for consumer goods. Heat setting of glass fabrics imparts good wrinkle resistance to fabrics made from these fibres. Glass fabrics have low moisture absorbency and good hand and luster when used in home furnishing applications.

The high density of glass provides reasonable draping properties for most applications, although glass fabric constructions are extremely heavy in weight. Its sunlight and heat resistance make glass fibres useful in window coverings. Glass fibres are extremely difficult to dye, and pigmentbinder or pigment-melt colouration must be used.

Glass substrate must be laundered or dry-cleaned using minimum mechanical action and drip drying. Glass substrates do not need ironing, although careful ironing up to 400°C is possible. Glass and related inorganic fibres are the only truly non-flammable man-made fibres. They are used extensively in curtains and draperies, electrical and thermal insulation, tire cord, reinforced plastics, industrial filters, and protective clothing and accessories. Apparel applications are limited to very fine Beta glass fibres.

## INORGANIC FIBRES

A series of man-made inorganic fibres other than glass exist that are non-flammable, heat stable amorphous materials useful in industrial fabric constructions, including refractory materials. These inorganic fibres include pure silica, potassium titanates, aluminum borosilicates, and aluminum oxide-zirconium oxide polymers, Most of these fibres have high strength, are less susceptible to chemical attack than glass, and melt above 1000°C. They may be used in higher-temperature applications than possible with most glass fibres.

## ASBESTOS

Asbestos is the name given to several natural minerals which occur in a fibrous crystalline form. The asbestos is initially crushed to open up the fibre mass, followed by carding and spinning to yield fibres of circular cross section 1-30 cm in length. Asbestos is very resistant to heat and burning, to acids and alkalies, and to other chemicals. Although it has low strength, asbestos fibre does not deteriorate in normal usage, and it is not attacked by insects or microorganisms. Asbestos is used in fireproof clothing, conveyor belts, brake linings, gaskets, industrial packings, electrical windings, insulations, and soundproofing materials. Inhaled asbestos fibres have been shown to be a serious health hazard, and it has been removed from the textiles market.

## METALLIC FIBRES

Metallic fibres are defined as fibres composed of metal, plasticcoated metal, or metal-coated plastic. Single-component metallic fibres for textile usage are fine drawn filaments of metal which can be spun and woven on normal textile machinery. These metallic fibres possess the properties of the metal from which they are formed. Multicomponent metallic fibres are more commonly used in textiles and are usually made from flat aluminum filaments surrounded with or bonded between clear layers of polyester, cellophane, or cellulose ester or from polyester film which has been metallised through vacuum deposition of aluminum and then encapsulated in polyester. In general, the properties of these fibres resemble the properties of the plastic film used to form the multi component fibre. The fibres are generally weak and easily stretched but can be used for decorative purposes and for applications where electrical conductivity and heat resistance are important. Trade names for metallic fibres include Brunsmet and Lurex.

## MISCELLANEOUS FIBRES

In this stage, fibres which do not logically "fit" under other classifications are listed—novaloid, carbon, poly-.!!!-phenylenedibenzimidazole and polyimide fibres. These fibres were developed for specific industrial applications and do not find wide use in consumer goods. For the present, it appears unlikely that new generic fibre types will be introduced in the marketplace unless they have distinguishing characteristics that make them uniquely suited to particular applications. All of these fibres possess such properties.

## NOVALOID

Novaloid is the designation assigned by the FTC for a class of flame retardant fibres made from crosslinked phenol-formaldehyde polymer. The fibres of this class in U.S., production are called Kynol and are manufactured by American Kynol, Inc. The fibre is golden yellow in colour and possesses good physical and chemical properties.

The fibre is thought to be formed through spinning phenol-formaldehyde prepolymer, followed by heating the fibre in formaldehyde vapor to crosslink the structure. The fibre has low crystallinity with a tenacity of 1.5-2.5 g/d and an elongation at break of about 35 per cent. The fibre has a specific gravity of

1.25 and a moisture regain of 4 per cent-8 per cent. It is inert to acids and organic solvents but is more susceptible to attack by bases. The fibre is totally heat resistant up to 150°. This characteristic yellow fibre is dyeable with disperse and cationic dyes. Novaloids are characteristically used in flame retardant protective clothing and in apparel and home furnishings applications where low fibre flammability is desired.

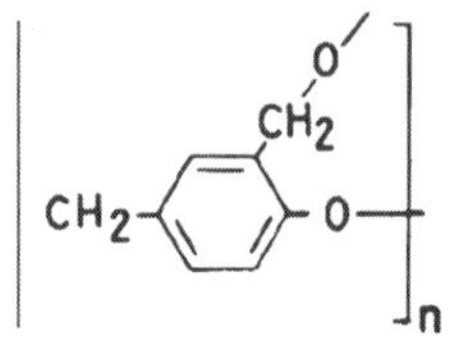

NOVALOID

## CARBON

Carbon or graphite fibres have been developed recently for use in industrial and aerospace applications. The carbon fibres are prepared from rayon, acrylic, or pitch fibres by controlled oxidation under tension in limited oxygen atmosphere at 300°-400°C. At this stage of oxidation, carbon fibres have sufficient flexibility to be used in apparel applications such as flame retardant clothing.

Further oxidation at temperatures near or exceeding 1000°C under tension results in a high strength fibre consisting of a continuous condensed aromatic network similar to graphite. Fully oxidised carbon fibres have tenacities from 10 to 23 g/d dry or wet and elongations at break of just 0.4 per cent to 1.5 per cent. They have specific gravities from 1.77 to 1.96 and are mildly electrically conducting.

Although the fibres are of high strength, they exhibit poor abrasion characteristics and must be sized with epoxy or other resins before formulation into textile substrates. Carbon fibres are inert to all known organic solvents and to attack by acids and bases under normal conditions. These black fibres have excellent resistance to sunlight and biological agents, are inherently flame retardant and are resistant to oxidation at high temperatures.

*Carbon fibres are extensively used in reinforcing fibres for resins and plastics in high performance fibre-polymer composites and are marketed as:*

- Celion,
- Hi-Tex, and
- Thornel.

## POLY-M-PHENYLENEDIBENZIMIDAZOLE (PBI)

PBI was developed by the U.S. Air Force and Celanese as a flame retardant fibre for use in aerospace applications. The fibre is spun from N'N.

dimethylacetamide followed by derivatisation with sulfuric acid to form a golden fibre. It has moderate strength and good elongation at break, a moderate density, and a high moisture regain. Fabrics from the fibre possess good hand and drape and are stable to attack by ultraviolet light. PBI does suffer from poor dyeability, however. In addition to its low flammability, the fibre has a high degree of chemical and oxidation resistance, does not show appreciable heat shrinkage up to 600°C and generates only small quantities of smoke and toxic gases on ignition. In recent years, it has shown potential as a replacement for asbestos, as a flue gas filter material, and as an apparel fabric in specialised applications.

**POL YIMIDE**

An aromatic polyimide has been introduced by Upjohn Company for use in flame retardant, high-temperature applications. The fibre is spun from the polymer by wet or dry processing techniques using a polar organic solvent such as N,N-dimethylformamide to give a fibre with a round or dog-bone cross section. The highly coloured fibre may be crimped at 325°C using fibre relaxation. The fibre has a tenacity of 2-3 g/d and a 28 per cent-35 per cent elongation at break. The fibre has a moisture regain of 2.0 per cent-3.0 per cent and melts at about 600°C.

# 4

# Modernisation in Textile Industry

As a whole the textile industry has made major progress process modernization and automation over the last 5 to 10 years. The focus of these programs has been in one of many directions in various companies. Generally new technology has been utilized to reduce the number of process steps, but just as often to reduce the labour content of the process. This is especially evident in the reduction of low skill-high drudgery jobs, but it also applies to the reduction of human intrusion into textile processes, which inevitably is a quality argument.

Examples of these process improvements include the automation of opening rooms, the setting up of chute-feeds and high production cards, the automation, at least in part, of drawing, the introduction of open end spinning, the increasing use of shuttleless looms, the use of automatic systems for handling waste, and the nearly universal use of microprocessor-controlled monitoring and reporting of production variables. Activities in applying existing available technology to textile processes are being pursued in all of the major process categories.

## BALE STORAGE

There is potential for the automatic analysis of incoming bales of fiber, especially if a robot or other automatic device could be involved in the analysis and then direct the bale to the appropriate storage area.

Automated bale storage is a labour saving technology that leads to less handling as well as presenting the opportunity of coding each bale for future identification.

This is important due to the desirability of tracking the identity of the material in each textile process process from the initial bale through all of the process steps.

This in turn permits one to know the accurate history of all material and all processes and allows the solution of quality control problems on a more rational basis. Cards can be loaded and unloaded automatically and their output can be conveyed automatically to draw frames.

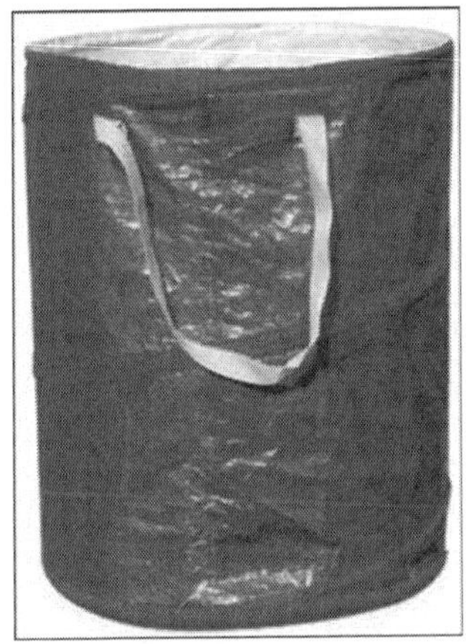

**Fig.** Bale Storage

However the requirement to mix the inputs to draw frames in such a way as to insure good blending is potent; these patterns also need to be capable of being changed from time to time.

**Fig.** Bale Storage Shed

**Preparationof Yarn**

The issues in yarn formation generally point toward reducing the number of steps wherever feasible. One view is that devising processes that can successfully do this is more potent than devising automated materials handling systems to move materials between currently popular process steps. Generally the ideal yarn manufacture process would take fiber from a bale, convert it to a sliver and move it to a spinning process, for example open end spinning, with automatic transfer of the yarn output either to a warper or to a loom.

Stress also must be placed on computer monitoring both of quality and of production rate, with the quality monitoring being tied to feedback and control mechanisms throughout the many processes. The monitoring technology is already available for drawing and the technology for monitoring either open-end yarn or other types of yarn is on the horizon. Another area of considerable importance with respect to monitoring is to be able to determine the need for machine maintenance by continuous monitoring of yarn from, for instance, open end rotors.

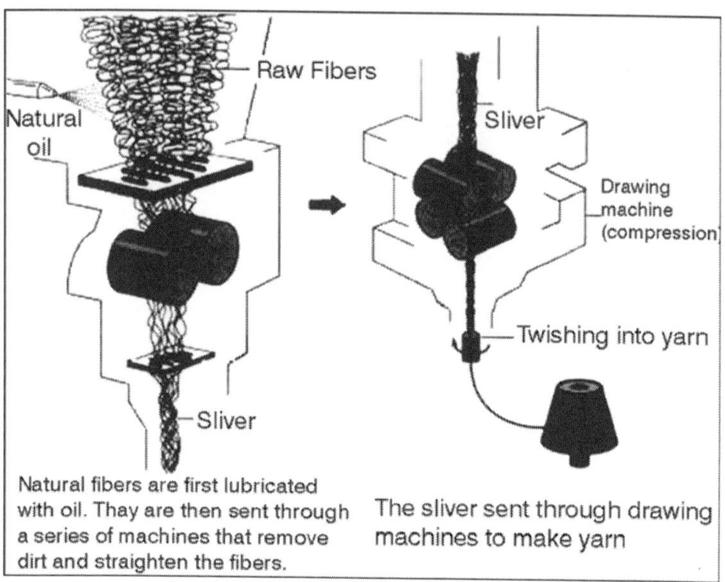


**Fig.** Preparationof Yarn

This needs that all ends be monitored regularly and that faulty positions be identified soon. The repair, when it is needed, could be made automatically or, alternatively, that end could be stopped and machine maintenance ordered. A useful system would also monitor and record for long-term gradual deterioration as opposed to short-term problems. In this whole system, the philosophy is to prevent defective material at its source. In general it is important to measure and to control quality on the spot at every step of the yarn formation process.

The computer technology is already available; the real problem seems to be the development of appropriate sensing elements. There are a number of opportunities in present yarn processes for automated materials handling, including the use of robots.

This is especially true for systems that have reduced the number of process steps, i.e., open end or other systems, but it could also be utilised with ring spinning technology in which the automation of roving frames is a current need. Connecting winders to large spinning frames is a good idea, and it was demonstrated at the 1983 ITMA, but winders require to be designed to accommodate some flexibility of yarn count; this perhaps could be accomplished by designing winders with space for extra positions. This linking technique permits less handling of yarn packages, which is a distinct advantage, and it also allows better package identification and control. Although techniques for the continuous monitoring of different stages of yarn processes are available, knowing where to direct that information or what correclive actions to take still is nat know to a large extent.

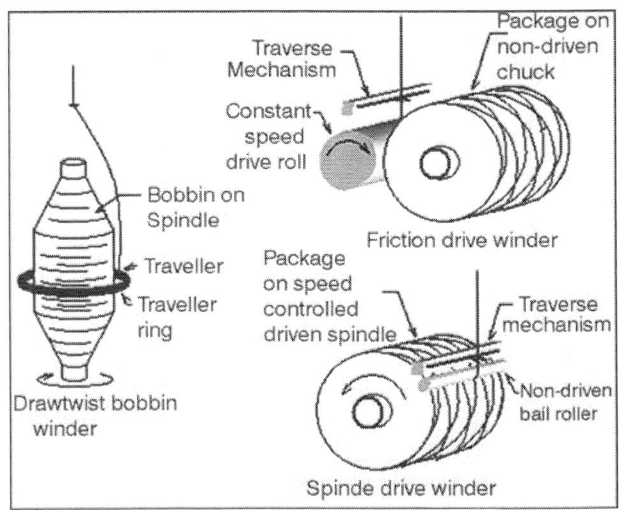


**Fig.** Different Stages of Yarn Processes

Since it is necessary to be able to identify abnormal parts of the process, the real issue is to condition where in a process faults occur, what their relations with other process steps will be, and the feedback loops in which to exercise control.

## PREPARATION OF FABRIC

The issues in fabric preparation are almost all in the areas of monitoring and control, inspection, and material handling. Modern weaving and knitting machinery allows the choice of several options and a conscious decision of whether to stress productivity or flexibility. The important trend, outside of the design of fabric formation machinery, is the elimination of menial jobs that mostly are related to materials handling.

As an instance, fabric handling with driverless tractors is well-proven now, although this is an area in which potential savings other than labour savings are nat easy to identify; therefore this may remain fairly standard technology for the future. It also would be desirable to devise systems for automatic doffing of cloth and probably automatic cutting, in conjunction with automatic inspection and grading.

Inspection of fabric is thought by several to be a process which should be capable of elimination, if proper process monitoring and control systems can be invented. There is, however, a difference in philosophy depending on whether a greige mill or a dyeing and finishing plant is involved. In the former case it is perhaps more likely that inspection could be eliminated if it were to be carried out later on the finished fabric. Perhaps there will always be a need for some inspection, but it should be automated. In partical it should be capable of viewing the fabric with both the plant specifications and the customer's specifications in mind, changing the latter as the customer changes and changing the former as the fabric style changes from lot to lot. These issues become more significant as the textile industry as a whole moves to higher quality and output levels as then inspection speeds become the limiting process speeds. Computer-controlled inspection systems also ought to interface with computer-controlled cutting systems to optimize the cutting of first quality yarn lage.

Monitoring in weave rooms specially, but also in other fabric formation areas, is done today almost solely for the objective of generating management information. Although this is an important function of such systems, what is needed, besides, is diagnostic monitoring that will locate and diagnose loom malfunctions, preferably before they lead to producing off-quality material. Such systems would be part of a larger-monitoring system that deals with the flow of raw material into the fabric formation process as well as with fabric formation itself. The development of truly reliable monitoring and control systems is closely association with the development of automated inspection systems, since the latter rely on the assurance that defects are minimized in the process.

## Dyeing and Finishing

### *Discreate Process*

In this area there is a common view that the number of discrete processes in dyeing and finishing needs to be reduced drastically, and as many as feasible should be combined. The goal would be to make dyeing and finishing a truly regular process rather than a series of batch processes, each with its own control and materials handling problems. What this implies is the development of some rather sophisticated monitoring and control systems for dyeing and finishing processes generally. Among these systems, the more potent seem to be the ability to monitor and control the colour of wet fabric, as well as to monitor and control moisture content.

**Fig.** Dyeing Fabrics

The objective is to be able to predict accurately the colour of the final dry fabric by measuring its characteristics at the moment it is being dyed.

Continuous dye ranges now have significant automation but are hampered by their inability to run very small lots efficiently. Systems need to be developed that will permit rapid change over from lot to lot on a continuous range system with a minimum fabric band between the changes. In addition to suitable monitoring and control functions, ways to rapidly alter dye baths in order to change colour fastly must be developed.

The aim is to produce systems with very rapid response times and this will require precision instrumentation for adding chemicals and controlling the parameters of the process. For similar reasons it is extremely important to have absolutely uniform desizing and bleaching. In this area there my well be applications also for computer-based monitoring, feedback, and control systems. There also are general needs for reducing the energy cost in dyeing by reducing either the setting or the drying needs. And for some products, it would be very useful to have dyeing the last of all finishing steps in order to improve order and warehouse versatility.

## PROCESS OF AUTOMATION IN HANDLING OF MATERIALS

From raw materials to the finished product there exist major opportunities for process automation in materials handling. Current applications of robotics are nearly all in this area, for the reason that it seems comparatively easy either to identify labour savings or to accomplish a reduction in menial and sometimes hard to fill jobs related to these processes. A significant concept that accompanies automated material handling is automated identification of the material which, if carried through in all parts of the process permitt the identification of quality control problems regarding their material source.

Much automated materials handling is, or will be, built into the process technology itself; but there remain materials handling problems that will need special technologies, the newest of which are driverless vehicles and robots. On the other hand, it can be considered that automated materials handling is really a second choice to the integration of processes; to the extent that some processes cannot be integrated, materials handling will remain a essential technology. There is much technology applications development to be accomplished, in the robotics area, which is not yet, in the view of some, suited for most textile applications. Particular issues comprise combining carding with breaker drawing with no intermediate materials handling, the doffing of ring spinning frames, the doffing of cloth at the loom, and the creeling of packages for dyeing, finishing, and warping.

### Manufacturing Process of Apparel

It is easiest to approach this topic by beginning at the front end of an apparel manufacturing process, so the first topic should be computer-aided design. Unfortunately, in its apparel design context, this is a technology that is not quite ready for commercialization. It has the same goal as computer aided fabric design systems, namely, permitting the designer to quickly, and easily test his or her ideas at a computer console with an appropriate viewing mechanism to see the result of those ideas.

The problem in apparel design is inevitably that of converting easily and reliably between two-dimensional and three-dimensional information. Technically it would appear easier to take the two-dimensional information from a set of pieces (for instances, for a shirt) and change them to a three-dimensional image of the shirt's final appearance.

However, designers often operate in the reverse mode, that is, they make an illustration of a garment the way they realise it, and then the problem is to convert that picture to a series of two-dimensional pieces to outline the actual manufacturing process. Research progresses and so many theoretical reports are available in this area but, as far as commercialization of this technology goes, progress remains slow.

However, the situation is significantly better in the area of computer-aided marker-making, grading, and cut-order planning. There are many systems on the market that permit an experienced operator to take a standard pattern piece, generate other pieces from it on a graded size basis (i.e., grading) and then to pack all of the necessary pieces on a piece of fabric so as to give maximum fabric utilization. Maximizing fabric use is perhaps one of the greatest driving forces for this technology.

It is estimated that, for plain fabric, fabric wastage can be deducted to about 3 per cent with this technology compared with 5-6 per cent when human beings do the same job unaided. On the other hand, if instead of plain fabrics, one uses patterned fabrics in his process, then fabric wastage can be as high as 25 *per cent* and there is explicit room for improvement here. Although there is no current use of this idea, it appear logical that computer-aided pattern packing ought to be combined with fabric inspection information, performed at the textile manufacturer's plant.

Such a combination would allow the apparel maker to cut around any defects that might be present and guarantee that those pieces that were cut were defect free; this would effectively increase fabric utilization. Technologies for cutting fabric are increasingly being debated on a more philosophical basis. The issues are whether, in fact, the conventional bundle system of fabric cutting and material transport can be retained if other parts of the apparel manufacturing system are to be highly automated. The bundle system uses a spreading machine (which some say has not seen any major improvement in several years) to lay and then cut as several as 100 layers of fabric at once. Thus, this system produces 100 layers of shirt fronts, 100 layers of collars, and 100 layers of pockets and these all are bundled together to give material at each sewing station. The merit of the system is that it is, or at least is thought to be, a very effective and cost efficient way to cut fabric; therefore, it has inherently low costs related to with it. This assumes that the person who does the cutting, if it is done manually, doesn't make a mistake. One mistake, of course, can ruin 100 shirt fronts. More importantly deducted, from an industrial engineering stand point, is that the bundle system inherently provides a buffer for each of the other steps in the process. If a sewing machine breaks down or if an operator is sick one day, the manufacturing process can still proceed because of the inherent build up of material available at all stations. The chief demerit of the bundle system is the long in-process time of the item being gathered. It is said that for some articles of clothing the in-plant time is as high as 20-30 days.

This represents an enormous inventory cost and is considered by some to be a major detriment to productivity enhancement. The bundle system also assumes that fabric is identical from one layer to another as the identity of several pieces effectively is lost in the system. To the extent that fabric is not uniform, then there are inevitable wastages because fabric pieces cannot be properly matched. The alternate system, used in some parts of the apparel industry where fabric is extremely expensive and a cutting mistake cannot be risked, is, for the sake of a better term, the one-layer system.

As the name means, each piece is cut individually in this system and then the problem is to route the pieces in the correct order to their assembly points. The merit of this system, other than the fact that the penalty for cutting errors is minimized, is that this process is actually more amenable to automation and control of the rest of the assembly process.

Besides, such a cutting system permits an enormous reduction of in-process time from 20-30 days to, in some cases, perhaps a few hours. It has obvious advantages. The major disadvantage of the one-layer system is that there is no inherent buffer capability. If a piece is missing or if an assembly station is inoperative, then the whole production process is held up.

On the other hand, it may be feasible to attack this problem by considering the "just in time" manufacturing philosophy that automobile makers use, in which a forward part of the manufacturing process calls to it material just as it is required and no sooner. There is no buffer concept in that philosophy, but it needs adequate material handling systems and reliable parts interchangeability. There are major innovations, some of them now quite old, in the area of cutting fabric.

Computer-controlled cutting, as exemplified by the Gerber cutter, is a technology that has been around for probably 10 years. It is precious machine and is utilized only by large companies that can justify having large cutting departments. In its present form, it takes information directly from the computer-aided marker-maker and cuts pieces from layers of fabric in a very fast and reproducible manner. It also uses a reciprocating knife in most instances, as this has, up to now, been the standard cutting technique. There are more novel cutting techniques available, each of which has been downgraded in the past because of the demerit they have in working with the bundle system. Laser cutting is a technology that is used to cut out men's suits, for instance.

It can only work really well on one or a few layers of fabric at a time, and it works best on non-thermoplastic fabrics that otherwise would melt at the edges of the cut. (The problem in working with several layers of fabric is that the cut edges tend to fuse together during the cutting process and then they are very difficult to separate.) Water-jet cutting is a technology used in many other fields, such as the shoe industry, the automobile industry, and in a number of industries where complex shapes have to be cut rather rapidly.

Water jets don't work well with the bundle system either, as the power in the jet is severely retarded after the first few layers of fabric. Not with standing, if the one-layer system were to be adopted more uniformly, then either laser cutting or water-jet cutting systems would make considerable sense. Especially this is true as either of these technologies can be combined with robotic guidance of the cutting head. Joining techniques will continue, most people think, to emphasize sewn seams. Consumers hope certain aesthetic and mechanical attributes of sewn seams, and these are difficult to achieve with any other technology. Modern sewing machines operate mechanically much like the one that How invented, but they have advanced to the point where microprocessor control is an inherent part of several of them.

Microprocessors can control both the stitch pattern and the feeding and ejecting of material once it has been sewn. An emerging technology, not commercial yet, is three-dimensional sewing, in which a sewing head on a robot arm effectively operates in three dimensions on pieces of fabrics that have been placed on a form. There is a considerable amount of research going on in this area. In the sphere of non-sewn techniques, both adhesives and thermal bonding techniques are important.

Again, there seem to be applications for robot systems in this area, although the number of seams that can use these-technologies perhaps is limited. As in the textile industry, materials handling in the apparel industry is a major problem, which can be divided into many aspects. The first is materials presentation at the sewing machine itself. Some apparel companies ascribe as much as 80 per cent of their labour cost to the sewing part of their operations but closer inspection divnlges that actual sewing only contributes perhaps 20 per cent of this cost; the rest is materials handling. There are so many of devices already available to bring material in a more orderly fashion to the sewing machine operator. Two such are made by Gerber and by Eaton. In each case, a rather complex overhead track system is loaded with pieces and partly assembled systems and moved by operator demand to the place where sewing is to take place.

These systems appear to be logical components of more automatic systems for the future. Several have investigated robots or robot-like systems as devices for materials handling, but the chief drawback is the classic limp fabric problem. It seems that it is nearly impossible to design a machine to pick up a piece of fabric, which may have only one of so many weights and stiffnesses, and to operate on it in a reliable and consistent manner. There exist devices which can handle certain classes of fabrics.

The Clupicker is definitely one of these, and there now exist robotic end effectors that are claimed to be able to operate with a wide variety of fabric types. One of these is comprised in the Draper project technology. Nevertheless, until the problems of reliable' materials flow, not only at sewing stations but between sewing stations, is solved, automation in any real sense apparel manufacturing will be severely handicapped. The manufacturing loop requires to be closed with sufficient management information, not only to track a process but also, at least in the ideal case, to determine when a process is reaching trouble.

There are a number of management information packages being sold into the apparel industry now which rely for the most part on operator input. Considering that a sewing machine operator has so many other things to do, one has to wonder just what the quality of the input data to these systems really is. Nevertheless, this represents a significant improvement from the traditional information gathering technology that has been prevalent.

## INCREASING USE OF AUTOMATION

The general themes of all of these comments for all three industries are the increasing use of computer technologies, the requirement for the invention of suitable monitoring, feedback, and control loops and the increasing use of automation in materials handling. These trends arise because processes of the future in these industries will most desirably have minimum numbers of intermediate steps and minimum numbers of people involved in these steps.

Another general aim is to design processes with less dependence of unit cost on volume, which therefore would permit management more choice of flexibility or mass throughout for any special product at any particular time. In the fiber industry the will be on developing high-speed extrusion processes that can be fully automatic. Very large packages will be produced of at least 100 pounds, which will mean less inspection. These packages will be loaded directly onto a creel either for warping or directly for weaving or knitting. Boxes and other temporary storage devices will no longer be used. In staple fiber production the fiber instead of being processed into bales, may well be packed directly into a truck and on delivery, transferred to card chute feeds. In the textile industry, there are those who predict common automation of processes, especially materials transfer among carding, drawing, spinning, and fabric formation processes. Special technology developments may include new spinning technologies that would either replace or refine current open end technology; these will become essential because cotton production is producing increasingly trashier material. Many respondents suggest that replacing the card is a necessary feature of future technologies, largely because cards don't permit for easy material transfer to the next step. An ideal system would send fiber directly to open-end spinning.

If cards are not replaced, provision for loading and unloading them automatically will require to be made and, in traditional systems, sliver cans will be transferred automatically to draw frames. The output from these can be transferred automatically to open end spinning frames and robots may become increasingly useful at this stage. In fabric production, an ideal situation will permit changing styles on looms much more rapidly than at present, and it will also include the rapid transfer of computer-generated fabric designs to looms for trial runs. Fabric grading will be largely automatic with the ability to distinguish not only sorts of defects, but also their placement within the fabric and to characterise those conditions with reference to either plant specifications or customer specifications.

Such systems will work on any style and any colour of fabric. Dyeing and finishing, but especially dyeing techniques, will be entirely automatic, capable of making nearly instantaneous changes from one shade to another and capable of reproducing shades absolutely. The implications for computer monitoring, feedback, and control of these systems are so many. Management information will be much more complete and easy to use than it is today. The information collecting used for managerial decision making will be part of the same process monitoring and control information gathering that will be essential to run an error-free process. There also will be a real requirement to be able to estimate cost, labour and material savings, and management time in the process of determining whether a new device or technology is worthwhile at a particular cost.

The emphasis, in general, in the textile industry as well as in the fiber industry will be on upstream quality improvement, which is more likely to produce a higher quality product than automated inspection systems. In the apparel industry, stress will be placed on more efficient materials use, and on speeding up material flow through the factory.

This may be more important in the long term than direct labour savings. However, the cost of inventory is so high that any technology that allowed a garment to pass through the system in a few hours as opposed to several days would be very cost-effective. There will be increasing emphasis on automation in sewing and other joining technologies and material transfer from station to station. This is the thrust of both of the widely publicized research programs in the automation of apparel manufacturing: the Draper Laboratories programme and the Japanese MITI programme. Both of these programs use the concept of the "manufacturing cell", in which robots, hard automation, computer-aided design, and computer-driven information flow combine to define a self-contained manufacturing unit. The first is an American Project sponsored by the Tailored Clothing Technology Corporation, abbreviated (TC)2" but loosely called the "Draper" project after the Draper Laboratories where the work actually is being carried out. The philosophy of this project is that a manufacturing cell exists wherever a sewn seam must be accomplished, and that the significant problems are getting material into the system, performing some orientation and/or parts identification, probably folding the material, sewing it in a reliable fashion, and ejecting the material from the system.

The Draper prototype machine that was completed some time last year (reputed in its present form to cost around $50,000) consists of a number of novel devices to do these things. The important technology appears to many to be in the use of vision systems to identify and orient fabric parts, and in the claimed invention of a robot that can, in fact, reliably pick up single layers of fabric (although in this case it is for folding; i.e., moving one layer over another) and in a sewing head that travels to the place where the seam is to be made.

Each of these operations is microprocessor controlled and, at this time, the process has been designed to do only one thing, yo sew the sleeves of men's coats. The Draper project investigators claim that the technology is adaptable to a wide variety of apparel assembly processes and some of its aspects certainly are, especially the vision system and the materials handling technology. Some in the apparel industry criticize the' project, partly on the ground that it represents a very costly, very specialized machine, and partly on the basis that nothing has been done to change the sewing technology at all.

A project receiving almost equal publicity in this country is a Japanese project in which a group of Japanese companies under the auspices of the Japanese government (MITI) have banded together to form an organization known as Technology Research Association of Automated Sewing Systems, which has published a very elaborate programme for the entire automation of apparel manufacture—what really amounts to a completely automated factory.

Whereas the Draper project has cost on the order of $5 million so far, the Japanese project is reputed eventually to cost $50 or $60 million, and there are several who believe that it will cost much more than this before it is finished. At this stage, planning and some fundamental work are being done in the areas of sewing preparation, sewing and assembly, fabric handling, and production control. Thus, all of the apparel manufacturing process is being attacked, with the desired end result being an automated factory composed of so many individual work cells, into each of which material is fed as appropriate and taken out as appropriate, with the entire process operating very much as an integrated system. While the Draper project emphasis is partly on the reduction of labour content, the Japanese stress is almost completely on the elimination of labour content. Several explain this as the difficulty in Japan of obtaining apparel plant labour, and suggest that this really is an effort to preserve an industry for which a labour force has largely ceased to exist.

**The Hereafter**

Are we approaching the day when textiles and apparel will be assembled in an automated factory? This is the objective of several manufacturing industries.

The traditional definition of the "factory of the future" is one that combines computer-aided design, hard automation, robotics, and computerized information systems to give a self-contained manufacturing cell that has as its input computer generated information and has as its output both computer generated information and a product that is uniformly manufactured to a specific quality standard. Whether or not this all can happen in the textile industries is dependent on other, nontechnical, aspects of these industries.

If any of these levels of automation are to take place, they almost certainly will take place according to company type and, as a result, several predict that the overall textile industry will become divided into two basic sorts of companies, those that are more manufacturing-oriented and those that are more marketing-oriented. The marketing-oriented companies will tend to be design-oriented, specialist companies who place foot market response above production efficiency. They will react very quickly to the marketplace, they will manufacture high value-added products, and they will use less of the available technology than they might if they were manufacturing larger amounts of their product. However, some technologies will be potent to them, particularly in the area of computer-aided design and in management information systems.

They will tend to judge automation by its abilily to enhance the value that they consider most important, i.e., flexibility regarding market demands. On the other hand, manufacturing-oriented companies will be very large and they will make very large amounts of their product. They will be prime candidates for large scale automation and perhaps even for the automatic factory. At any rate, they will have to have the ability for great investment of capital and this; as much as anything, will condition whether a company can afford to be manufacturing-oriented or not. The primary thrusts of these companies will be productivity and quality and, with a view to realise these, they will effort very hard to lower their overall labour costs (not only their direct labour costs) and to lower their in-process time so as to escape the high cost of inventory.

## CLOTHING MANUFACTURING

The energy consumption share of the clothing manufacturing division which consists of large numbers of small-sized companies and their employees in the overall textile industry is not necessarily low but the ratio of energy cost to the total cost is relatively low. However, the energy cost forecast is inevitably a gradual increase under circumstances where the production of high value-added goods is required, along with the implementation of labour saving measures, as a result of the challenging market environment characterized by personalized and diversified consumer needs, high demand for quality goods, short product cycles, etc. Therefore, it is desirable that a comprehensive rationalization programme be investigated apart from reductions in energy consumption.

### TEXTILE INDUSTRY

In order to promote energy conservation in the textile industry, this manual analyses in detail the actual conditions of energy consumption in each field. Thus, it has been compiled to serve as reference material for the practical application of these techniques which are always kept available for the engineers. A step-by-step description is included as to how to implement the energy conservation measures, specifically in the area of the dyeing and finishing process, where many small- to medium-sized companies are operating. We strongly hope that this manual will be used as a guide to promote energy conservation in the textile industry and to rationalize the management.

### STRUCTURES OF TEXTILE MARKETS

In general, it is well known that the developmental process of a textile market moves from a product-oriented stage, where all goods produced are sold to a consumer-oriented stage where the desired goods are only those which will satisfy the consumer's demand, comprising his wants and purchasing power. This is via a mass-consumption stage which is supported by a mass-production system.

The characteristics of textile products demanded in the markets in these different developmental stages may be classified into price-sensitive mass-production type basic clothing, which applies to the earlier stages, and multi-line, small-volume production type fashion clothing which pertains to the last stage and needs to satisfy each individual consumer's taste.

### BASIC CLOTHING

Assuming the amount of fabric necessary for people to live a physically comfortable life in a given environment can be derived from the weight of clothes which is enough to keep a certain level of body temperature, given such factors as air temperature, humidity and wind velocity prevalent in the region, basic clothing can be regarded as a cumulative total of such clothes.

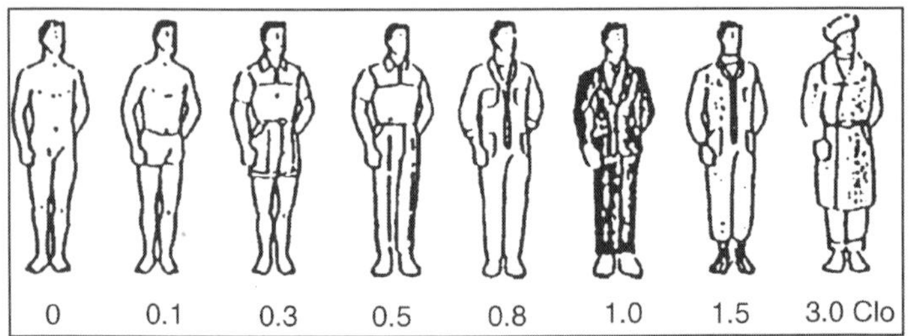


**Fig.** Typical Relationship between Clothing and Clo Value

Figure Summarizes the Relationship between Some Clo Values and Examples of the Corresponding Clothes. Table. shows the results of calculations to find the weight of clothes required under average climatic conditions in summer and winter in three Japanese cities.

**Table. Example of Effect of Climatic Conditions on Clo Value and Weight of Clothing**

| Region | Season | Temperature (°C) | Humidity (%) | Wind Velocity (m/sec) | Clo Value | Required Weight of Clothing (g) |
|---|---|---|---|---|---|---|
| Sapporo | Jan | –4.9 | 74 | 2.3 | 4.1 | 4005 |
| | Aug | 21.3 | 79 | 2.7 | 1.2 | 1189 |
| Tokyo | Jan | 4.7 | 53 | 3.1 | 3.1 | 3034 |
| | Aug | 26.7 | 75 | 3.0 | 0.6 | 607 |
| Fukuoka | Jan | 5.7 | 67 | 3.7 | 3.0 | 2937 |
| | Aug | 27.3 | 76 | 2.9 | 0.6 | 607 |

*Notes*:

1. The clo values are shown assuming an average skin temperature of 33.5°C.
2. One unit of clo value is 0.155°C/W/$m^2$.
3. The required weight values are calculated with Hanada's formula for female clothing:
   - Y = 0.00103W - 0.025
   - Where Y: Clo Value
   - W: Total Weight of Clothing

Although the ultimately required quantity is difficult to determine, the per capita annual textile consumption (kg/person) can be estimated through multiplying the above weight by coefficient a (obtained from the service life of a textile product and its number of units demanded per annum).

## CHARACTERISTICS OF TEXTILE MARKET

Most textile consumption takes place in apparel products, and the general tendency is that where further expansion is intended, it will be carried out through the development of industrial applications in which textile products are used as production materials (ranging from fishing nets, tire cords and canvas cloth to geotextile). Therefore, in order to study the textile manufacturing industry, it is generally sufficient to consider the production of apparel goods.

## WORLD'S TOTAL TEXTILE DEMAND AND PRODUCTION BASE DISTRIBUTION

Figure shows the relationship between world population and the total textile demand.

Assuming a global environment in which world population will grow from the present 5.4 billion to 10 billion in 2050, and further to 11.6 billion in 2150 when it is expected to reach a static state, the total textile consumption is forecast to double, even using the current figure of per capita annual average textile consumption (8kg/person).

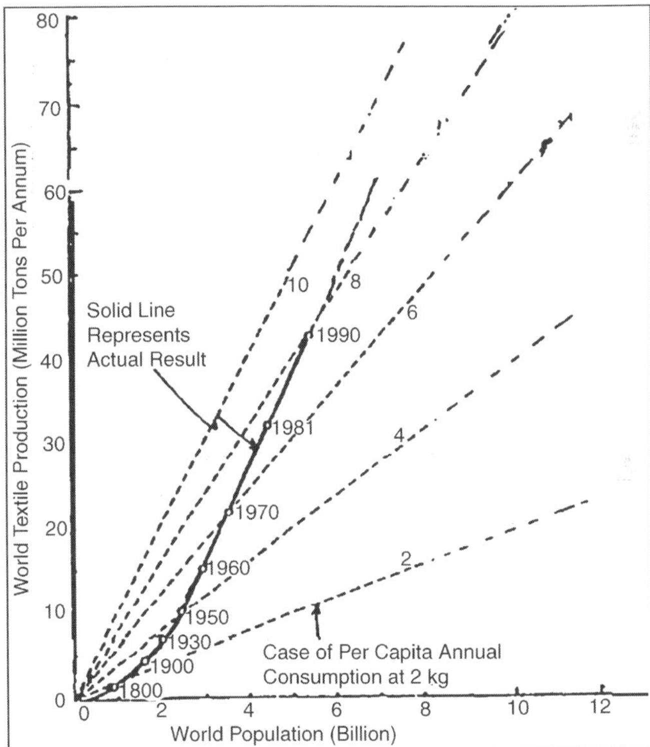


**Fig.** Relationship between World Population and World Textile Consumption and Production

The textile industry is traditionally regarded as a typical labour intensive industry developed on the basis of abundant labour supply and has a tendency to expand to overseas markets once the domestic demand is satisfied, as illustrated by examples of the established textile industries of many developed countries. For this reason, even when the textile industry of a specific country is to be examined, it is widely recognized that a business strategy taking into consideration the global textile industry setup is very important. Figure shows an example of the schematic representation of such a global setup of textile industries.

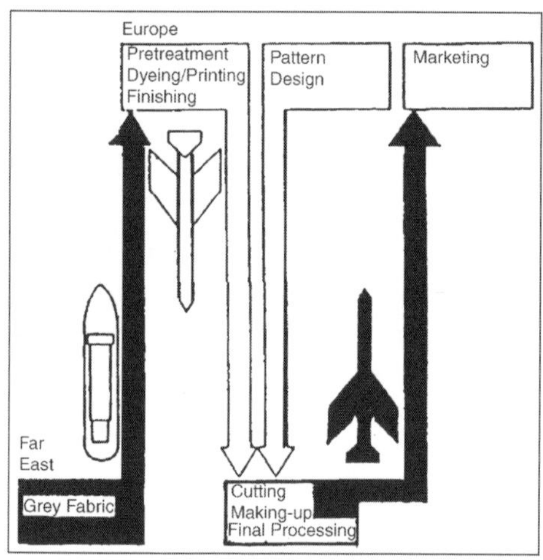


**Fig.** Example of Schematic Representation of World Textile Industry

**Specialized Techniques Necessary for Apparel Goods Production**

There are a number of intertwined technical factors involved in various subprocesses or specialized technical fields which make up the overall production process that fabricates and reshapes the raw fibre material into the final textile product to be used by the consumer.

## COMPARISON OF CHARACTERISTICS OF SPECIALIZED TECHNICAL FIELDS

In Fiscal Year 1989, the textile industry had a share of 4.4% in shipment value. 14.8% in the number of businesses (20%), and 10.4% in the number of employees (23.3%) in the overall Japanese manufacturing industries, characterized by a gradual decline in relative share, although in terms of absolute shipment value it has actually expanded to 1,053% of the Fiscal 1955 level.

As has been frequently pointed out throughout its development, the Japanese textile industry has a unique organizational structure consisting of groups of independent companies where all companies in a group belong to one of the specialized technical fields and operate in a horizontal specialization configuration. A comparison and summary of the major indices of these company groups, classifying them in accordance with their specialized technical fields. From these indices, common company characteristics for each group may emerge.

Namely, in terms of business size, the fibre production and spinning subdivisions are in contrast with the rest of the textile industry where relatively large numbers of small businesses coexist. In addition, these groups of small companies can only stay in business by relying on the supply of abundant cheap labour, exhibiting a legacy of the textile industry's past as a labour intensive industry, even on its path towards modernization. Table compares the energy consumption shares of various specialized technical fields and it can be seen that energy consumption is relatively high in the fields of dyeing and finishing, fibre production, spinning, weaving and clothing manufacturing.

As for water consumption, the share of the textile industry in the entire manufacturing industries is 5.2% for fresh water and 1.1% for sea water. Considering the fact that most of this water is used for easy-to-recycle temperature control and cooling purposes, the industry's total water consumption cannot necessarily be regarded as high. However, as is widely accepted, the dyeing and finishing division is placed in a special position in that its water consumption is mainly for processing and washing purposes.

## THE GLOBAL TEXTILE AND CLOTHING INDUSTRIES

Textile making is a very ancient craft, with a history almost as old as mankind itself. Remembered and recorded in poetry and ancient stories and myths, textiles have always been important to man. As well as providing protection from the elements, the first textiles were used as decoration, providing status for the owner. They were also used as tools; bags for transporting belongings and for holding food as it was gathered. Textiles are produced in almost every country of the world, sometimes for consumption exclusively in the country of manufacture, sometimes mainly for export.

From cottage industry to multi-national corporation, textiles and clothing are truly global industries. In 1782, the invention of the steam engine gave the world a new power source and started the Industrial Revolution. Previous to this the production of textiles had been a domestic system, a cottage industry with textiles spun, knitted and woven in the home. By the middle of the nineteenth century, however, there was a whole range of new machines and inventions that were to take textiles into an era of mass production in factories.

The development of man-made fibres and new dyestuffs in the early part of the twentieth century, and continuing technological developments, have led and continue to lead to new products and applications. The actual processes of textile manufacture, however, are still very much as they have always been, with the vast majority of cloth being woven or knitted from yarn spun from fibre. And, while much production may be very technologically advanced, hand-produced textiles are still made in many countries exactly as they were many, many years ago. Nowadays, many different types of companies are involved in the production of textiles and clothing world-wide; some companies own many huge manufacturing plants in many different countries while others will have only a few employees and some may not actually manufacture at all.

## TEXTILE MATERIALS, PROCESSES, AND PRODUCTS

Fibres are manufactured or processed into yarns, and yarns are made into fabrics. Fabrics may be manufactured by a variety of processes including knitting, weaving, lace making, felt-making, knotting, and stitch bonding. These fabrics may be industrial textiles with detailed technical and performance specifications, or they may be sold either to retail or contract as apparel, furnishings or household textiles, where aesthetics may be as, or sometimes even more important than performance.

The fabrics may be coloured by dyeing or printing, or be finished to enhance their appearance or performance. A wide diversity of products are made from textile products or have some textile components; textiles go into car tyres, and geotextiles are used for lining reservoirs, while medical applications include artificial ligaments and replacement arteries.

### Design in textiles and Clothing

Every textile product is designed: that is, it is made specifically to some kind of plan. Design decisions are made at every stage in the manufacturing process—what fibres should be used in a yarn, what yarns in a fabric, what weight of fabric should be produced, what colours should the yarn or fabric be produced in, what fabric structures should be used and what finishes applied. These decisions may be made by engineers and technologists in the case of industrial or medical textiles where performance requirements are paramount, or, more often in the case of apparel, furnishings and household textiles, by designers trained in aesthetics, technology and marketing. The designers found in the textile and clothing industries are frequently involved throughout the design process, from initial identification of a need/requirement, through research, generation of initial design ideas, design development and testing to ultimate product specification.

### Designers Found in the textiles and Clothing Industries

*The designers found in textiles and clothing include*:

- Colourists predicting and forecasting future colour ranges
- Yarn designers
- Knitted fabric designers
- Woven fabric designers
- Carpet designers
- Print designers
- Embroidery designers
- Knitwear designers

- Garment designers
- Accessory designers
- Print producers
- Stylists
- Colourists developing colourways
- Repeat artists

**Fibres**

Fabric is made from yarn, and yarn is made from fibres. These fibres can be either natural or man-made. Natural fibres include animal fibres (*e.g.* wool and silk), vegetable fibres (*e.g.* jute and cotton) and mineral fibres (*e.g.* asbestos). Man-made fibres are either regenerated or synthetic; viscose rayon, based on regenerated cellulose, is man-made but not synthetic while polyester, polypropylene and nylon are all synthetic fibres. Synthetic fibres are produced by the large chemical companies including Dupont, Bayer, Hoechst and Astra Zeneca. Many of these companies produce no fabric but specialise in the production of certain types of fibre which they sell on as fibres or manufacture into yarns.

**Yarns**

Yarn producers or spinners buy in natural and/or man-made fibres to make these into yarns of different sizes and characters; regular and fancy yarns. For many years the main spinning systems could be given as woollen, worsted and cotton, and these systems gave rise to the woollen, worsted and cotton industries. Developments in spinning, however, have led to new spinning systems including 'open-end', 'self-twist' and 'jet' spinning.

At its simplest, yarn production is essentially about taking fibres, organising them so that they lie in a lengthways direction and twisting them to create a yarn. By combining fibre types, and using different spinning systems and machinery, yarns can be developed with individual profiles suitable for a vast range of end uses. Regular yarns are those which have a regular straight profile and these can be twisted together, making 'two-fold' or 'three-fold' yarns for example. Fancy yarns can be created by deliberately introducing irregularities or intermittent effects along their length. Yarns can be combined together as components of new yarns with different effects and properties from their component parts. As well as changing the appearance of a fabric, the introduction of a fancy yarn will affect the handle and performance of that fabric.

**Woven Fabrics**

Strictly speaking, the definition of a textile is 'a woven fabric' but the term textile is now considered to cover any product that uses textile materials or is made by textile processes. Essentially, woven fabrics are structures produced by interlacing two sets of threads; the warp which runs in a lengthways direction and the weft which runs in a widthways direction. Weaving methods include tapestry and jacquard.

**Knitted Fabrics**

Knitted fabrics are produced by interlacing loops of yarn. In weft knitting, loops are formed one at a time in a weft-ways direction as the fabric is formed. Hand-knitting with a pair of knitting needles is weft knitting. In warp knitting there is a set of warp yarns which are simultaneously formed into loops. To connect these chains of loops the warp threads are moved sideways in such a way as to cause the loops to interlink.

**Lace and Non-woven Fabrics**

Fabrics may also be produced by methods other than weaving and knitting. Lace is an open-work fabric made by looping, plaiting or twisting threads by means of a needle or a set of bobbins. Fabrics produced by crochet and macramé are often called lace, although strictly speaking they are not. Knotting is another way of making fabrics. Knotting was a popular pastime for women in eighteenth-century Europe and colonial North America, and one method still seen today is macramé.

A knotting process is also used for fishing nets, and some rugs and carpets are knotted — made by tying yarns onto a foundation weave. There is also a group of fabrics called non-wovens which include true felt (where animal fibres are matted together) and fabrics produced by bonding webs of fibres together by stitching or by sticking with adhesive. However, in terms of volume produced, knitted and woven fabrics are by far the most common methods of fabric production.

**Fabric Terms**

A length of woven or knitted fabric is usually referred to as a 'piece'. Often, fabric woven by a mill will not be coloured and this undyed fabric is called 'grey cloth'. Colour can be added by dyeing the piece, and such fabric is referred to as being 'piecedyed'.

Colour can also be added to a fabric by applying pigments or dyes in a printing or other colouring process after weaving or knitting, or by using already dyed yarns in the construction of the fabric. Cloth made from dyed yarns will not normally be dyed again or printed. 'Finishing' is what happens after the fabric has been made. The finishing processes employed will be determined by the type of fabric and its performance requirements. Any excess dye will normally be removed, any applied pigment will normally be set, and any dye will be fixed. Fabrics may be brushed or raised to enhance appearance and handle, or fire-retardant and soil-resist treatments may be applied. Fire retardancy may be a product performance prerequisite; anti-soiling and anti-static finishes, while not necessarily pre-requisites, enhance performance, as do methods of coating fabrics to produce microporous surfaces.

### Geography and Fabric Types

Certain countries, and areas within countries, have developed industries around specific fibres/fabric types and there are still parts of the world where craftsmen produce fabrics exclusive to them, such as some hand-crafted batiks and weaves. In the UK, Manchester was nicknamed 'Cottonopolis' as it was built in the main (as was much of Lancashire) from money earned through the cotton trade. Scotland developed a woollen trade through both woven and knitted fabrics, while, again in the UK, Yorkshire was home to the worsted industry and the Midlands became famous for knitting and lace making. India has a history of cotton manufacture and, in the eighteenth century, was famous throughout Europe for its mordanted cottons or chintzes.

## RECENT TRENDS IN TEXTILE MANUFACTURE

Throughout the 20th century, the trend in textiles has been toward lighter-weight materials. Better transportation and improved indoor heating have made warmth a less important aspect of clothing for most people than it was in earlier centuries. Since 1900 the weight of average clothing fabric has fallen by more than one-third. When people need extremely warm clothes, they tend to wear special outer garments such as parkas that today are typically made of lightweight synthetic materials.

Synthetic materials, produced from chemical compounds rather than from plant fibres or animal hair, have provided less-expensive substitutes for natural fabrics. Synthetic materials can also be superior to natural fibres in strength and durability.

Rayon was an early synthetic substitute for silk. Nylon, a synthetic fabric introduced in the 1930s, was another early substitute for silk and quickly became the fibre of choice for women's stockings. Polyester, a form of plastic, was introduced in clothing in the early 1950s. Blended with rayon or cotton, polyester found its first use in so-called wash-and-wear fabrics that needed little or no ironing.

Synthetic fibres fell out of favour in the late 1960s but new kinds of polyester that are more durable and have a softer, more natural feel to them have become increasingly popular in the late 20th century. Synthetic fibres such as spandex have revolutionized clothing by making possible the production of extremely flexible, form-fitting garments. Other synthetic fibres, created for special purposes, range from lightweight but extremely warm or water-resistant fabrics, such as polypropylene and the composite polymer Gore-tex, to woven, bullet-proof fabrics such as Kevlar that serve as body armor.

## HAT FELT

HAT, a covering for the head worn by both sexes, and distinguished from the cap or bonnet by the possession of a brim. The word in O.E. is *hcet,* which is cognate with O. Frisian *hatt,* O.N. *hotte,* &c., meaning head-covering, hood; it is distantly related to the hood, which is cognate with the German for "hat," *Hut.* The history of the hat as part of the apparel of both sexes, with the various changes in shape which it has undergone, is treated in the article Costume. Hats were originally made by the process of felting, and as tradition ascribed the discovery of that very ancient operation to St Clement, he was assumed as the patron saint of the craft. At the present day the trade is divided into two distinct classes.

The first and most ancient is concerned with the manufacture of felt hats, and the second has to do with the recent but now most extensive and important manufacture of silk or dress hats. In addition to these there is the important manufacture of straw or plaited hats and hats are occasionally manufactured of materials and by processes not included under any of these heads, but such manufactures do not take a large or permanent position in the industry.

*Felt Hats:* There is a great range in the quality of felt hats: the finer and more expensive qualities are made entirely of fur; for commoner qualities a mixture of fur and wool is used; and for the cheapest kinds wool alone is employed. The processes and apparatus necessary for making hats of fur differ also from those required in the case of woollen bodies; and in large manufactories machinery is now generally employed for operations which at no distant date were entirely manual. An outline of the operations by which the old beaver hat was made will give an idea of the manual processes in making a fur napped hat, and the apparatus and mechanical processes employed in making ordinary hard and soft felts will afterwards be noticed.

Hatters' fur consists principally of the hair of rabbits (technically called coneys) and hares, with some proportion of nutria, musquash and beavers' hair; and generally any parings and cuttings from furriers are also used. Furs intended for felting are deprived of their long coarse hairs, after which they are treated with a solution of nitrate of mercury, an operation called carroting or *secretage,* whereby the felting properties of the fur are greatly increased. The fur is then cut by hand or machine from the ski and in this state it is delivered to the hat maker.

The old process of making a beaver hat was as follows. The materials of a proper beaver consisted, for the body or foundation, of rabbits' fur, and for the nap, of beaver fur, although the beaver was often mixed with or supplanted by a more common fur. In preparing the fur plate, the hatter weighed out a sufficient quantity of rabbit fur for a single hat, and spread it out and combined it by the operation of bowing. The bow or stang ABC was about 7 ft. long, and it stretched a single cord of catgut which the workman vibrated by means of a wooden pin E, furnished with a half knob at each end. Holding the bow in his left hand, and the pin in his right, he caused the vibrating string to come in contact with the heap of tangled fur, which did not cover a space greater than that of the hand.

At each vibration some of the filaments started up to the height of a few inches, and fell away from the mass, a little to the right of the bow, their excursions being restrained by a concave frame of wicker work called the basket. One half of the material was first operated on, and by bowing and gathering, or a patting use of the basket, the stuff was loosely matted, about 50 by 36 in., called a bat. In this formation care was taken to work about two-thirds of the fur down towards what was intended for the brim, and this having been effected, greater density was induced by gentle pressure with the basket.

It was then covered with a wettish linen cloth, upon which was laid the hardening skin, a piece of dry half-tanned horse hide. On this the workman pressed until the stuff adhered closely to the damp cloth, in which it was then doubled up, freely pressed with the hand, and laid aside. By this process, called basoning, the bat became compactly felted and thinned toward the sides and point.

The other half of the fur was next subjected to precisely the same processes, after which a cone-shaped slip of stiff paper was laid on its surface, and the sides of the bat were folded over its edges to its form and size. It was then laid paper-side downward upon the first bat, which was now replaced on the hurdle, and its edges were transversely doubled over the introverted side-lays of the second bat, thus giving equal thickness to the whole body. In this condition it was reintroduced between folds of damp linen cloth, and again hardened, so as to unite the two halves, the knitting together of which was quickly effected. The paper was then withdrawn, and the body in the form of a large cone removed to the plank or battery room.

The battery consisted of an open iron boiler or kettle A filled with scalding hot water, with shelves, B, C, partly of mahogany and partly of lead, slop ing down to it. Here the body was first dipped in the water, and then withdrawn to the plank to cool and drain, when it was unfolded, rolled gently with a pin tapering towards the ends, turned, and worked in every direction, to toughen and shrink it, and at the same time prevent adhesion of its sides.

Stopping or thickening any thin spots seen on looking through the body, was carefully performed by dabbing on additional stuff in succes sive supplies from the hot liquor with a brush frequently dipped into the kettle, until the body was shrunk sufficiently (about one-half) and thoroughly equalized. When quite dried, stiffening was effected with a brush dipped into a thin varnish of shellac, and rubbed into the body, the surface intended for the inside having much more laid on it than the outer, while the brim was made to absorb many times the quantity applied to any other part.

On being again dried, the body was ready to be covered with a nap of beaver hair. For this, in inferior qualities, the hair of the otter, nutria or other fine fur was sometimes substituted. The requisite quantity of one or other of these was taken and mixed with a proportion of cotton, and the whole was bowed up into a thin uniform lap. The cotton merely served to give sufficient body to the material to enable the workman to handle the lap.

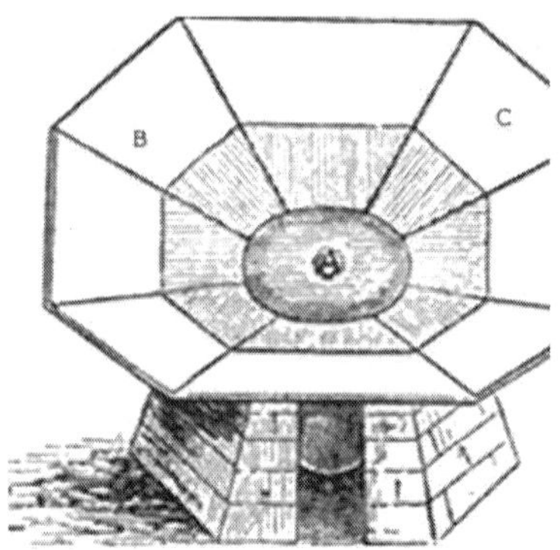


**Fig.** Hat Felt Machine

Being damped, the workman spread over it a covering of this lap, and by moistening and gentle patting with a brush the cut ends of the hair penetrated and fixed themselves in the felt body. The hat was then put into a coarse hair cloth, dipped and rolled in the hot liquor until the fur was quite worked in, the cotton being left on the surface loose and ready for removal.

The blocking, dyeing and finishing processes in the case of beaver hats were similar to those employed for ordinary felts, except that greater care and dexterity were required on the part of the workmen, and further that the coarse hairs or kemps which might be in the fur were cut off by shaving the surface with a razor. The nap also had to be laid in one direction, smoothed and rendered glossy by repeated wettings, ironings and brushings. A hat so finished was very durable and much more light, cool and easy-fitting to the head than the silk hat which has now so largely superseded it.

The first efficient machinery for making felt hats was devised from the United States the machine-making processes were introduced into England about the year 1858; and now in all large establishments machinery such as that alluded to below is employed. For the forming of hat bodies two kinds of machine are used, according as the material employed is fur or wool. In the case of fur, the essential portion of the apparatus is a "former," consisting of a metal cone of the size and form of the body or bat to be made, perforated all over with small holes. The cone is made to revolve on its axis slowly over an orifice under which there is a powerful fan, which maintains a strong inward draught of air through the holes in the cone. At the side of the cone, and with an opening towards it, is a trunk or box from which the fur to be made into a hat is thrown out by the rapid revolution of a brushlike cylinder, and as the cloud of separate hairs is expelled from the trunk, the current of air being sucked through the cone carries the fibres to it and causes them to cling closely to its surface. Thus a coating of loose fibres is accumulated on the copper cone, and these are kept in position only by the exhaust at work under it.

When sufficient for a hat body has been deposited, it is damped and a cloth is wrapped round it; then an outer cone is slipped over it and the whole is removed for felting, while another copper cone is placed in position for continuing the work. The fur is next felted by being rolled and pressed, these operations being performed partly by hand and partly by machine.

In the case of wool hats the hat or body is prepared by first carding in a modified form of carding machine. The wool is divided into two separate slivers as delivered from the cards, and these are wound simultaneously on a double conical block of wood mounted and geared to revolve slowly with a reciprocating horizontal motion, so that there is a continual crossing and recrossing of the wool as the sliver is wound around the cone.

This diagonal arrangement of the sliver is an essential feature in the apparatus, as thereby the strength of the finished felt is made equal in every direction; and when strained in the blocking the texture yields in a uniform manner without rupture. The wool wound on the double block forms the material of two hats, which are separated by cutting around the median or base line, and slipping each half off at its own end. Into each cone of wool or bat an "inlayer" is now placed to prevent the inside from matting, after which they are folded in cloths, and placed over a perforated iron plate through which steam is blown.

When well moistened and heated, they are placed between boards, and subjected to a rubbing action sufficient to harden them for bearing the subsequent strong planking or felting operations. The planking of wool hats is generally done by machine, in some cases a form of fulling mill being used; but in all forms the agencies are heat, moisture, pressure, rubbing and turning.

When by thorough felting the hat bodies of any kind have been reduced to dense leathery cones about one-half the size of the original bat, they are dried, and, if hard felts are to be made, the bodies are at this stage hardened or stiffened with a varnish of shellac. Next follows the operations of blocking, in which the felt for the first time assumes approximately the form it is ultimately to possess.

For this purpose the conical body is softened in boiling water, and forcibly drawn over and over a hat-shaped wooden block. The operation of dyeing next follows, and the finishing processes include shaping on a block, over which crown and brim receive ultimately their accurate form, and pouncing or pumicing, which consists of smoothing the surface with fine emery paper, the hat being for this purpose mounted on a rapidly revolving block. The trimmer finally binds the outer brim and inserts the lining, after which the brim may be given more or less of a curl or turn over according to prevailing fashion.

### Silk Hats

The silk hat, which has now become co-extensive with civilization, is an article of comparatively recent introduction. It was invented in Florence about 1760, but it was more than half a century before it was worn to any great extent. A silk hat consists of a light stiff body covered with a plush of silk, the manufacture of which in a brilliant glossy condition is the most important element in the industry. Originally the bodies were made of felt and various other materials, but now calico is chiefly used. The calico is first stiffened with a varnish of shellac, and then cut into pieces sufficient for crown, side and brim. The side-piece is wound round a wooden hat block, and its edges are joined by hot ironing, and the crown-piece is put on and similarly attached to the side. The brim, consisting of three thicknesses of calico cemented together, is now slipped over and brought to its position, and thereafter a second side-piece and another crown are cemented on.

The whole of the body, thus prepared, now receives a coat of size, and subsequently it is varnished over, and thus it is ready for the operation of covering. In covering this body, the under brim, generally of merino, is first attached, then the upper brim, and lastly the crown and side sewn together are drawn over. All these by hot ironing and stretching are drawn smooth and tight, and as the varnish of the body softens with the heat, body and cover adhere all over to each other without wrinkle or pucker. Dressing and polishing by means of damping, brushing and ironing, come next, after which the hat is "velured" in a revolving machine by the application of haircloth and velvet velures, which cleans the nap and gives it a smooth and glossy surface. The brim has only then to be bound, the linings inserted, and the brim finally curled, when the hat is ready for use.

## RUGS AND CARPETS

Rugs and Carpets, heavy fabrics commonly made of wool and used as floor coverings. A rug differs from a carpet in that it is usually woven in one piece and can be of any size but usually does not cover an entire floor.The term *rug* is derived from the Scandinavian word *rugga* by way of the old Norwegian word *rogg,* which meant a wool covering for the bed or body. For several centuries in Europe, the term *rug* denoted a rough, heavy woolen fabric characterized by a coarse, napped finish and used as apparel by the poorer classes.

The term *carpet* was used originally to describe coverings for tables, beds, and other furniture, and only from the early 18th century was it associated with the floor. The modern usage is imprecise and includes all woven floor coverings and some textiles, such as wall hangings, furniture coverings, and saddlebags, made with a knotted pile or woven like a tapestry. The word *carpet* is ultimately derived from the Latin *carpere,* "to pluck or seize," thus implying a plucking of wool or carding of wool fibres, and reflects the fact that for centuries wool has been used in making carpets.

## FLAT-SURFACED WEAVES

The earliest carpets, made before the first millennium BC, were flat-surfaced weaves.

### Tapestry

Tapestry is probably the oldest of the flat-surfaced patterned carpet weaves. It is characterized by wefts, or horizontal threads, that do not run the full width of the fabric. Instead, discontinuous wefts of different colours form the design patterns.

Tapestry looms vary in form from culture to culture, although in general they have two stationary parallel bars and one or two adjustable bars around which the warp (vertical threads) is secured. They may be either vertical or horizontal, with heavy beams between which the warp is stretched. Auxiliary bars are attached by means of loops, or heddles, to alternate warps and enable the weaver to pass weft threads between odd and even warps. The warp-weighted loom, a type of upright tapestry loom, was used in ancient Greece and remains in use in Scandinavia. It is so named because the warp hangs free from the upper bar, weighted at the ground with ceramic or stone weights tied to the end of the warp.

### Soumak

Soumak, another technique for making flat-woven rugs, originated in the Middle East as early as the 7th century BC; fine contemporary specimens are made in Caucasia. Wefts are wrapped onto the warps in a weft-wise (lateral) direction. Frequently, rows of soumak are alternated with rows of plain weave.

### Others

Other flat-woven rug techniques include brocading (introducing supplementary wefts or warps to create a pattern) and embroidery. In some cases, two or more techniques may be used in the production of a single rug.

## KNOTTED-PILE WEAVES

The characteristic feature of pile carpets is the knotting and fastening of extra strands of material into a foundation weave to form a dense layer of loops that completely covers the foundation on one side. Short individual strands are manually wrapped or knotted around taut warp threads. After each row is knotted, a weft thread is carried across the full width of the web and beaten firmly into place. The warp runs the length of the carpet, and the fineness of the weave depends on the thickness of the warps and their proximity to one another. Generally, looms for weaving knotted-pile carpets are the same as those for tapestry.

## MATERIALS

The most common material for rugs and carpets has traditionally been sheep's wool, although in certain regions goat's or camel's hair is also used. Luxury carpets are woven with silk pile. Although wool is often used for warp, cotton is more common because of its smooth surface and its resistance to stretching. When gold and silver threads are used, they are generally brocaded into the woven surface.

### Natural Dyes

Until the late 19th century, natural dyes were used for colouring weaving yarns. Information about dyeing practices before the 16th century is limited, but it is known that all dye matter came from mineral pigments, from vegetal (plant) materials, or from animals and insects. Mineral pigments such as ocher (yellow or red), limestone or lime (white), manganese (black), cinnabar (red), blue azurite, green malachite, lead oxide (red), and lapis lazuli (blue) were probably the first materials used for dyeing. Vegetal dyes came from leaves, roots, bark, and, occasionally, fruit or flowers of plants. Woad, a plant of the mustard family, and indigo, a bush from the pea family, were both used for blue dye.

Four vegetal yellows were especially important—saffron, safflower, weld, and fustic. Madder and redwoods had been used since ancient times for orange colours, and browns and blacks came from logwood, betel nuts, walnut and butternut hulls, and a gum resin called cutch. Animal sources of dyes included shield scales or cochineal insects found on cacti in North and South America, and kermes, a scale insect found on oak trees near the Mediterranean; both produced reds and pinks.

## THE TEXTILE MACHINERY INDUSTRY

The emergence of specialized machinery producers was a distinguishing feature that differentiated the United States and Great Britain from other 19$^{th}$ century textile centers. Supported initially by the size of the domestic textile industry, such specialization made possible the extreme adaptation of technology to distinct national conditions in these two cases. Other countries, beginning later and relying on imported machinery, typically had to choose between one or the other of the two dominant national models. As Kristine Bruland has emphasized, late industrializing countries such as Norway did not just buy spinning machinery on the international market, but more commonly an entire "package" of ancillary services, including technological information and supplementary machines, often accompanied by expert advisors and even skilled laborers.12 As the century progressed and British firms developed expertise in ring spinning, countries were increasingly able to compromise, dividing their investments between rings and mules. But they still relied heavily on British advice in doing so. In both the U. K. and the U. S., machinery producers played an active, initiating role in disseminating new technology to new producing centers. American textile-machinery manufacturing had important linkages to other branches of the machine tools industry. As early as the 1830s, machine shops that were initially attached to textile mills began to diversify their product lines into steam engines, turbines, locomotives, and other machine tools.

In contrast to the bifurcation across national boundaries, a tendency towards standardization within the country was observed very early, promoted both by long-distance sales of specialty firms and by the high geographic mobility of 19$^{th}$ century mechanics.

Towards the end of the century, New England machinery firms actively promoted textile development in the southern states, offering discounted machinery prices, technical advice, and even investment capital at times.13 In Britain, specialized machinery producers also sprang up with the rise of Lancashire in the first half of the 19$^{th}$ century, but quickly adopted a more expansive and outward-looking posture than did the parent industry.14 These firms were among the leading advocates of lifting the mercantilist restrictions on machinery exports, and took full advantage of their opportunities when that effort succeeded in 1843.

The industry leader, Platt Brothers of Oldham, was the largest engineering firm in the world as of the 1850s, and foreign sales accounted for nearly two-thirds of its receipts over the entire period 1873-1913.15 The pioneering British ring producers were Samuel Brooks (1872) and Howard and Bullough (1878); but by the 1880s, Platt Bros. and other firms were producing a full range of rings, mules and ancillary machinery.

The Chairman's annual report to the stockholders of Platt Brothers for 1888 noted that the company was by far the largest producer of ring frames in the world, that its machines were unsurpassed for excellence and speed, and that they were scarcely able to keep up with demand.16 But all the major firms drew upon expertise accumulated over most of the 19$^{th}$ century; the only significant new entrant after 1870 was Tweedales and Smalley, in 1891. By 1913, British machine makers supplied 87 percent of world trade in spinning and preparatory machines.

## HYPOTHESES AND EVIDENCE

The business records of the major British textile machinery firms are now available at the Lancashire Public Records office in Preston. Over many years time, we have assembled what we believe to be the most complete data set available on production and sales of spinning machines by these firms, covering the years 1879 to 1933. During most of this era, Lancashire was the world's dominant supplier to every country outside of the United States. The records thus offer a rare opportunity to trace the evolution of world spinning technology across this entire period, not only as it was embedded in machines, but as it was implemented in culturally and geographically diverse parts of the world.18

What do we hope to learn from a review of the specifications of textile machinery across fifty years of history? At least three types of questions present themselves:

- Was there any positive technological progress in textile machinery? Technical or engineering-based studies of technological change in history are comparatively rare. Most economic studies try to infer "technological progress" from some form of productivity index, a procedure that frequently conflates technical change with other forms of economic adjustment. The late nineteenth-century textiles industry is frequently classified as "mature," not subject to additional rapid improvements; indeed, this maturity is often linked to Britain's more general problem of maintaining world industrial and technological leadership. Yet we know that textiles technology experienced explosive change both before 1880 and after 1913. Do we actually see a hiatus of thirty years or more, or do the data show more gradually emerging trends? If progress occurred, along what technical dimensions and with what implications for international competition?
- Did rates of technical improvement differ between rings and mules? We know that the pace of change between continuous and intermittent spinning was highly uneven across two hundred years. But was the gradual expansion of the ring relative to the mule between 1880 and 1913 driven by continuing unevenness in the pace of advancement? Such a finding would have interesting implications for our thinking about the nature of technological change and biases in the flow of technological progress. A full understanding of these issues calls for more than conventional measures of productivity. At any point during their coexistence, rings and mules maintained market niches based on their known relative advantages and disadvantages. The test of their relative progress, therefore, would involve an attempt to measure the boundaries of these market niches and to trace their movement over time.
- Were there significant differences between nations in their capacity to absorb advances in technology? The technical specifications of machinery purchased are not "pure" reflections of supply-side developments in machinery production. Such indicators as machine size and machine speed, for example, partly reflect advances in the machines themselves, but they also involve greater demands on the skill and effort-level of the textiles work force. Comparative studies show striking differences between national industries in levels and growth of productivity, even where the capital goods were coming

from the same handful of British companies. It has been plausibly argued that these differences arose from varying success in recruiting and retaining a good work force, and in mobilizing the labor force to supply greater effort. Indeed, it may be difficult to identify the source of technical improvements, between these two general types. But the chance to examine an array of international cases may be very helpful in distinguishing changes that were common to all countries, versus those that were only experienced by the most advanced or the most successful national industries. In this regard, our data may enable us to address two specific issues that have been debated by economic historians:

- British Productivity Growth. An early study by Jones found no productivity growth This conclusion has been disputed by Sandberg and on different grounds by Lazonick.in British textiles at all, between 1870 and 1913.19 20 In both cases, the results turn crucially on appropriate measures of product mix and quality, and on input mix and quality. Productivity measures in this context generally suffer from problems of aggregation bias and inadequate measurement of quality. A better alternative is to look for evidence of changes in input-output relationships within a broad spectrum of technically specified parameters.
- Japanese Exceptionalism. Among all the newly industrializing countries of this era, Japan stands out for having broken out from the pack of poor countries of the Third World. Japan's record in cotton textiles was distinctive too, in that the country made the switch from mules to rings very early and very completely, and in defiance of the rules of thumb then prevailing in the industry: ring spinning technology arose in the high-wage context of the United States, yet Japan was relatively labor abundant; and ring spinning was most successful where the cotton fibres were medium- to long-staple, yet Asian cottons tended to be short staple. In our earlier work, we cited contemporary sources to the effect that the key to the Japanese switch was an innovative reconfiguration of major components of the production package. The Japanese industry, we argued, was able to deploy its largely female labor force to the task of judiciously blending imported and domestic cottons, making it possible to operate ring spindles at speeds up to 10,000 rpm without having frequent yarn breakages undermine efficiency.21 If this account is correct, we ought to be able to find evidence for it in the relationships among the choice of rings, machine speeds, yarn count and fibre length in Japan.

## NEW EVIDENCE: MACHINE SIZE AND MACHINE SPEED

The finding of a protracted period of coevolution between ring and mule, culminating in the near-total triumph of the ring by 1920 (Britain, France and Germany being the last bastions of mule holdouts).22 During 1878-1883, no country purchased more ring frames than mules. Clearly there was a global trend toward rings thereafter, but the pace of this shift was by no means uniform. Indeed, the share of mules in the world total actually increased during 1899-1906, compared to the previous period. This surprising temporary reversal of the trend largely reflects the prominence of Lancashire itself in the Edwardian textile construction boom, but not entirely so; a parallel reversal may be detected for Alsace, Canada and Spain.

In 1907-1914, the mule's downward share continued, but even at that late date many countries (not just Great Britain) made large purchases of new mule spinning capacity, the most conspicuous being Austria, Canada, France, Germany, India and Russia. The mule shares for the Continental countries may exaggerate the persistence somewhat, if German-made machines were mainly rings. Even so, it is undeniable that purchases of new mules were substantial in many countries, down to the very eve of the Great War.

The growth of spindles per frame, for rings and for mules. If staffing ratios per frame were relatively fixed, the rise in spindles per frame is a form of increased labor productivity. Most notably, the tables do not show a decisive performance difference between the two types of machinery. Both ring and mule frames increased in size between 1878/83 and 1907/14, the global average increase for mules actually outpacing that for rings slightly, 192 added spindles versus. (The increase for rings was slightly higher in percentage terms, because mules were three to four times larger at the beginning of the period.)

Perhaps more notable than the global averages are the patterns for different countries. As one might expect for a skill-based technology, the international dispersion in machine size for mules was significantly higher than that for rings. Spindles per mule frame in Britain led the world in every decade, the only exceptions being anomalous observations for Canada in 1891-98 and Japan in 1907-14.

Similar conclusions emerge when we examine patterns of *change* in machine size: for rings, the country observations cluster narrowly around the average increase of 30 percent; for mules, increases range from virtually zero in India and Russia to better than 50 percent in Austria, Italy, and Belgium. These figures suggest that for countries that were well adapted to the mule, productivity growth was clearly possible with that technology. By contrast, the growth in spindles per ring spinning frame was more uniform among nations. And in contrast to her world leadership in mule size, British rings were actually smaller than the world average, for every time period from 1878/83 to 1907/14.

Average speeds in ring spinning were somewhat below the U. S. norm, as Copeland reported. But ring speeds did increase over time, from an average of 8,100 rpm in 1884/90 (the first period in which ring speeds are recorded) to 8,900 in 1907/14. However, the same was true for mules; in fact, the overall increase in machine speed for mules actually exceeded the increase for rings. In the majority of cases for which comparisons are possible, mules were faster than rings.

True, the increase in world average speed for mules came to an end in 1914. But it is interesting to note that those countries with the most persistent commitment to the mule (Britain, India, and Russia) were also countries where mule speeds outpaced those of rings. Again we find that dispersion in mule speeds was far higher than dispersion in ring speeds. This contrast suggests that progress with mules was largely a matter of expertise and experience, whereas progress with rings had more to do with technical improvements in machine quality.

As measured by mule speed, Great Britain was the world leader in 1878/83. Thereafter, British speeds were sometimes matched by other countries, notably Russia. But Britain remained at or near the top in mule speed, while British ring speeds were not much different from the average. If we multiply Britain's 17 percent increase in mule size by her 20 percent increase in mule speed, the implied productivity increase is more than 40 percent over thirty to thirty-five years (a growth rate somewhat better then 1.0 percent per year). By no means is this figure a rigorous estimate of productivity growth; but one may also say that it is by no means symptomatic of a stagnant, unprogressive national industry.

## DIVIDING AND CONQUERING THE MARKETS: TRENDS IN YARN COUNT

The average yarn counts for which spinning machines were designed, pairing rings and mules by country. Whereas size and speed are open-ended performance characteristics, the distribution of yarn counts by machine type is a representation of the division of the product market between the two processes.

In Britain and some other countries, higher-count production was dominated by mules, rings entering mainly at the low end of the distribution. In that context, equal increases in yarn counts for rings and mules may not reflect 'neutrality,' or equal progress by both methods. A more plausible interpretation would be that rings were gradually broadening their commercially viable product space, while mules were retreating to a higher but smaller market niche. However, the overall distribution of counts in world markets was governed by a complex process that balanced relative costs against relative demands, interacting with uneven rates of technological change.

To add to the complication, the division of the market in other countries was often different. Because a chief advantage of the mule lay in its gentler and more flexible handling of fibers, it was sometimes observed that mules were preferred both for extremely fine spinning and for extremely coarse spinning — in the latter case, the true correlation being with the shortness of the cotton fibers used in coarse spinning.

Thus we find, for the earliest period in our sample (1878-1883), that rings in India, Austria, and Belgium were designed for counts *higher* than the average for mules. Thus, in the absence of a complete model of the cotton market, we have to interpret the count data with caution, focusing only on the most visible and persistent trends. The evidence does show an increase in average ring counts over time, but for the world as a whole, the extent of that increase was surprisingly limited prior to the 1920s: the global average was 25.9 in 1878/83, and had reached no higher than 29.8 by 1907/14.

One gets a different impression, however, by tracing the course of average ring counts in individual countries. Between 1878/83 and 1907/14, the average count for which new rings were designed increased by 20 percent or more in Britain, France, Italy, Russia, Spain, and Alsace, as well Japan and Mexico. Average counts for mules increased as well, but as just discussed, this probably reflected a decline in market share rather than an enhancement of the mule's productive range. Thus the country-by-country patterns do not appear entirely consistent with the aggregate.

There are several reasons why the global trend does not have to reflect the typical experience of individual countries. The average count for rings declined between the first and second periods, as Japan and India learned to adapt the ring to short-staple Asian cottons, substituting rings for mules in low-count production. Over a somewhat longer period, the increase in the global average ring count was held down by compositional change, specifically the growing market share of low-count producers – China as well as India and Japan – even while European countries were pushing the use of rings into the 30s and 40s for the first time. Thus the evidence is consistent with the view that a major frontier for ring-mule competition was an improvement in the ring's range of commercially viable counts.

What forces drove this process? Clearly one ongoing factor was the effort by machinery producers to extend their markets. The mule might match the ring in productivity growth for any given yarn count. But the mule's primary protection was its "preserve," the range of counts and qualities that a skilled mule spinner could achieve, beyond the reach of the ring at a point in time. Once the ring moved into new territory, matching productivity growth was not enough to save the mule, and competition in the machinery industry propelled advances precisely along these lines. In support of this view, we note that the $75^{th}$-percentile ring count (the count level below which 75 percent of the country's orders fell) increased in every case under study, an indication that intrinsic machine capacities were improving. As a 1909 observer put it, the self-acting mule was 'a beautiful piece of mechanism for performing one of the prettiest of operations;' but in modern mills, 'the roving is of such even thickness that ring yarn is practically perfect, and what more can mule yarn be?'

At the same time, the extension of the ring's domain was not purely a matter of technical progress in machine making; it also reflected the success of user countries in adapting their procedures and their labor force along this path. As noted, dispersion in performance measures using the ring was much lower than for the mule. But the data nonetheless provide support for a measure of Japanese exceptionalism.

After starting with mules in the 1870s, Japan's switch to rings was earlier and more complete than that of India, at a time when British and Indian opinion was distinctly divided on the merits of the two technologies. Alone among the emerging third-world textile countries, Japan quickly moved towards the top of the world distribution in the size and speed of its rings. And while still using a large proportion of short-staple Asian cottons as raw material, the average count of Japanese yarn advanced so rapidly that it was actually ahead of the world average as early as 1899-1906.

In contrast, China's textile industry remained concentrated on coarse yarn production throughout the 1920s. The performance problems of India's textile industry are well known, analyzed most recently by Wolcott and Clark. Whatever else the contrast between India and Japan may have represented, it did not derive from any difference in their access to world-class spinning machinery.

Thus it fell to Japan to "break the mold" of the pre-existing global division of the textile market, extending the reach of an American technology into a setting whose factor costs and market opportunities could not have been more different from those in the USA. Perhaps surprisingly, the closest parallel to the Japanese performance prior to 1914 is Mexico.

Despite its proximity to the United States, and despite a lively debate over the comparative merits of American versus British technology, the Mexican textile industry was supplied almost entirely by Lancashire ring spinning machines prior to 1918. Mexican mills were almost completely electrified by 1905, running their machines at speeds that matched the leading countries of the world. During the same period, average Mexican yarn counts increased even more rapidly than those of the Japanese, from 12.0 in 1878/83 to 27.8 in 1907/14. This record is confirmed by careful econometric studies, showing extraordinary rates of productivity growth in Mexican textiles through 1914.26 Thus Mexico provides another illustration of the democratizing potential of ring spinning, for countries with appropriate social and political characteristics, and access to growing markets. Unfortunately, the realization of that potential was interrupted by the Mexican Revolution, and the retreat to protectionism on both sides of the Rio Grande in the 1920s. The cross-section panel data provided by the records of the British textile machinery firms offers a new perspective on the dynamics of technological change and diffusion during an important half-century of world industrial history. Rather than seeing an older "mature" technology supplanted by a more advanced modern form, we observe two technological paradigms in competitive coexistence, each one capable of supporting ongoing productivity growth, through complementary improvements in machinery, organization and workforce skills.

The era saw the final phase of a longer competition between the principles of continuous and intermittent spinning. But behind this technical differences lay a deeper contrast between systems: a "British" craft-like technology, in which the machinery drew upon the personal skills of the operators, versus an 'American' approach in which improvements in the machinery reduced skill requirements and extended its range along other dimensions. At the same time, the diversity of experience among newly emerging industries makes it clear that progress was a two-sided affair, a mutual adaptation between machines and local conditions. Not all countries tapped into the ring's potential, and no other country matched Japan's success in adapting the 'American' principle to an entirely different economic and cultural environment.

From this vantage point, criticism of Lancashire's failure to switch more rapidly to the ring appears misguided. The mule was a skill-based technology, and in this competition, British mule spinners were the best in the world. Under machine-based ring technology, British productivity was not much better than the world average. Thus, it was *only* with the mule that the pioneer country could hope to retain its place in world markets. Once the mule itself ceased to be viable, no feasible choices could have staved.

## AUTOMATION IN TEXTILE DYEING INDUSTRY

Automation is concerned with the application of machines to task once performed by humans or, increasingly, to tasks that would otherwise be impossible. Although the term mechanization is often used to refer to the simple replacement of human labour by machines, automation generally implies the integration of machines into a self-governing system.

## USE OF COMPUTERS IN TEXTILE TESTING

### Automation of the Test Procedure

In such applications use is made of electronic processing power to control various aspects of the test rather than just to record the results. This means that steps such as setting ranges, speeds, tensions and zeroing the instrument can all be carried out without the intervention of an operator. The settings are usually derived from sample data entered at the keyboard. In the case of yarn-testing instruments can all be carried out without the intervention of an operator.

The settings are usually derived from sample data entered at the keyboard. In the case of yarn-testing instruments the automation can be carried as far as loading the specimen. This enables the machinery to be presented with a number of yarn packages and left to carry out the required number of tests on each package.

The automation of steps in the tests procedure enables an improvement to be made in the repeatability of test results owing to the reduction in operator intervention and a closer standardization of the test conditions. The precision of the instrument is then dependent on the quality of the sensors and the correctness of the sample data given to the machine. The accuracy of the results is, however, still dependent on the calibration of the instrument. This is a point that is easily overlooked in instruments with digital outputs as the numbers have lost their immediate connection with the physical world. If the machine fails in some way but is still giving a numerical output, the figures may still be accepted as being correct. The incorporation of computers and microprocessors has brought great changes to the instrumentation used for testing textiles. Their use falls into two main categories: recording and calculation of results and automation of the test procedure. Both of these uses may be found in the most advanced instruments.

**Recording of Results**

In these applications the computer is usually connected via an analogue to digital converter to an existing instrument from where it collects the data that would previously have been written down on paper by the operator.

*The advantages of such an installation are as follows*:

1. In the case of a graphical output the whole of the curve is recorded numerically so that results such as maxima, areas under the curve and slopes can be calculated directly without having to be read from a graph. This allows a more consistent measurement of features such as slopes which would previously have been measured by placing a rule on the graph by eye. However, it is important in such applications to be clear what criteria the computer is using to select turning points in the curve and at what point the slope is being measured. It is useful to have visual checks on these points in case the computer is making the wrong choice.
2. The ability to adjust the zero level for the instrument automatically. This can be done, for instance, by taking the quiescent output as being the zero level and subtracting this from all other readings.
3. The ability to perform all the intermediate calculations together with any statistical calculations in the case of multiple tests.
4. The ability to give a final neatly printed report which may be given directly to a customer.

It is important, however, to be aware of the fact that the precision of the basic instrument is unchanged and it depends on, among other things, the preparation and loading of the sample into the instrument by the operator and the setting of any instrumental parameters such as speed or range.

## FLEXIBLE AUTOMATION IN TEXTILE MANUFACTURING

In most sectors of textile manufacturing, automation is one major key to quality improvement and cost competitiveness. Early modernisation and technical developments in textiles concentrated on the automation of individual machines and their processes. Here, all process and machine variables were identified and placed under the surveillance of monitors or microprocessors. The machine and operating parameters of acceptable change were studied and programmed to control the quality and reproducibility of materials being produced. The next step involved the inter-linking of sequential machinery processes. Progress has been made in connecting operations such as yarn spinning systems, but considerable technical improvements are required to achieve fully automated textile mill operations.

A look at the 1990s gives evidence of where it is going. Throughout the 1990s, Computer Integrated Manufacturing (CIM) and Flexible Manufacturing Systems (FMS) have been the dominant production philosophies of the textile and clothing industries, both in developing and in developed countries. The ultimate goal seems to be the fully automated textile mill. On the whole, the industry has moved from the era of computer applications in textile operations to the era of computer integrated textile manufacturing.

The main objectives of Computer Integrated Manufacturing (CIM) are first, to provide accessible information for every sector of a plant for the efficient management of the various stages of production, second, to provide facilities for planning and control at strategic points, available for the directors, managers and supervisors to make decisions and third, to have compatible sophisticated high technology systems - particularly software - so that computers can talk to one another within the network, and modules can be linked with other modules, accepting additional work stations as the business grows.

## FIBRE MANUFACTURING

A major direction for evolutionary change in extrusion technology is the continuing integration of extrusion with downstream processes. Today, it is possible to find commercial examples for spin-draw-wind, spin-draw-warp and spin-draw-textile processes. These technologies place a different emphasis on material handling requirements; robotic technologies for package doffing and transport are increasingly available and yet because of the linking of processes, need be placed only at critical points in the overall process. The emphasis on flexible manufacturing, even in the fibre industry, has led to the development by some fibre producers of robotic techniques for the rapid change and replacement of spin packs and spinnerets. In these examples, robots are called upon to do what humans cannot do - change hot parts before they have cooled. Automated inspection of yarn packages for broken ends, poor package building, and improper tensions and misidentified packages is a goal being pursued by a number of fibre producers.

The history of the man-made fibre industry has emphasised process control more than any other segment of the textile operation. Increasing emphasis on product uniformity and adherence to quality standards continues to require fibre diameter monitoring, temperature and tension control, and monitoring of the solution properties of the polymer. These requirements are especially critical in micro-denier fibre extrusion, a process that produces fibres and eventually fabrics of truly different properties.

**Yarn Manufacturing**

Computer Integrated Manufacturing Systems are available that monitor and/or control practically all yarn production processes from opening and blending to spinning, winding and twisting. Applications include inventory control, order tracking, maintenance control, budgeting, mill management and many others.

Most companies now offer advanced controls on opening, blending, carding and other fibre preparation equipment, which are compatible with CIM. Ring spinning machines with individual spindle drives are available and these offer great flexibility and will readily fit into the CIM concept. Sliver weights can be controlled and the levels changed by on-machine electronics that can readily be connected to a computer network.

Online quality control in carding and drawing can perform spectral analysis and determine the cause of problems based on the frequency analysis of the defects. Yarn spinning is now so automated that a large spinning mill can be operated by a very small number of people since automatic end piecing and automatic doffing is performed by robotic mechanisms.

One of the world's most advanced examples of CIM applied to a spinning mill is Kondobo-Murata CIM mill at Horigane in Japan. The nucleus of Horigane plant is Murata's Link Coner Spinning/winding link system, while their 'Sky-Rev' automated inter-process transportation system operates between the post-carding sliver and ribbon-lapping and combing and again takes over to provide the automated transport link between combing and drawing.

**Fabric Manufacturing**

Weaving and knitting machine builders have been leading the way in utilising computer technology in textile manufacturing for many years with their use of CAD, bi-directional communication and artificial intelligence. With the availability of electronic dobby and jacquard heads, automatic pick finding, and needle selection, etc. these machines are the most easily integrated into computer networks of any production machines.

Bi-directional communication systems can be used to control many functions on a weaving machine. A CAD system can be used to develop the fabric to be produced and the design can then be transmitted over the network to the production machines to produce the desired fabric. Now, the design instructions can even be sent by modem from one country to a weaving machine located anywhere else in the world.

A weaving machine capable of receiving and responding to instructions in this way can therefore be operated in a developing country, while the designs it is weaving are originated and controlled, long-distance from a developed country. These technologies can greatly reduce the time needed to produce a fabric and give true meaning to the term 'quick response'. Weaving also is the area where artificial intelligence is progressing the fastest with developments such as expert systems to assist in troubleshooting looms.

In the 1990s, due to remarkable progress in computer technology, the application in sizing machines has increased to a greater extent such as multi-point thermo sensors for energy saving, automatic control of squeezing pressure, size pick-up detectors, multi-functional counters, etc. Sizing machine control systems provide a tool for management to insure that all warps are sized identically under standard operating conditions. These monitoring and control capabilities can be included in a computer network of a weaving mill.

For years knitting machine manufacturers have been making excellent use of electronics to provide machines that are more automatic and versatile and many refinements of these advances have been made. These automatic machines are already 'islands of automation' that can be incorporated into a CIM network.

Automated weaving plants are on the drawing boards. None is yet in operation but should be a reality within a few years. The six production steps winding, warping, sizing, weaving inspection and packing include 16 points of automation. Of these, 12 deal with materials handling or transport. Only four applications deal with automating the machine operations themselves. This includes automated process control on the slasher and the weaving functions of (1) Automatic Pick Repair (2) Automated Warp breakage Locator and (3) Computerised Machine Control. Manual assistance is still required for beam replacement and repair of warp breaks.

### Textile Dyeing

The automatic control of dyeing machines dates well back into the 1960s, and each succeeding year has shown miniaturisation and enhancement in the management of information on a timelier basis. The automation started with the introduction of a system that controlled a set temperature by switching heaters on or off. A short time later these were replaced by systems that controlled the dyeing cycle according to a time/temperature sequence. The processes of dye and auxiliary chemical addition as well as loading and unloading of textile materials were also automated to result in automated dye-house management.

A monitor displays scheduling for any machine and allows the operator to arrange the next lot. Batch weighing updates inventory each minute and give inventory of each dye by bulk and container. Any errors later in the process can be traced to a particular container if it should become necessary. Now, the jiggers have been fully computerised with total control over process. In the pad-batch dyeing system, the most outstanding development is special dye dispensing system, online colour monitoring and dye pickup control. The knowledge-based methods are becoming increasingly significant in the field of dyeing process automation.

Essentially Neural Network and Fuzzy Logic are frequently being used. The Glen Raven's new automated dye-house near Burlington, NC is among the most robotised plants in US textile industry. In the plant automated system directs the entire manufacturing process from dyeing to loading and unloading yarns.

It knows what colour and how much dye to add, when to mix it and when and where to route the yarn for the next step in the dye process. The system creates a highly effective and extremely efficient facility. Also, in India, a state-of-the-art indigo dyeing plant can be remotely operated and diagnosed for any maintenance across the globe.

## ONLINE QUALITY CONTROL

An important factor in the success of automated textile mill is online quality measuring, monitoring and controlling. More and more instrument companies have devices to perform these tasks while they were applied externally when introduced several years ago, today they are being incorporated internally.

The importance of online monitoring and quality control cannot be over emphasised. With the high rates of production now achievable, any off standard condition can produce large quantities of second grade material. This can represent non-recoverable value added production costs as well as the loss of full priced, first grade products.

Should the off-standard material remain in the production line, further deterioration in product quality such as, foreign-matter, broken filaments, slubs or unevenness can be expected in downstream processing. Additionally, machine stoppages can occur.

It is essential to incorporate online quality detectors that can measure quality on a continuing basis, adjust machine settings within prescribed tolerances to maintain nominal quality parameters, or stop production if automatic corrections cannot be made. Recent advances in imaging technology have resulted in inexpensive, high quality image acquisition and advances in computer technology allow image processing to be performed quickly and cheaply. This has given rise not only to a number of developments for laboratory quality testing equipments for fibres, yarns and fabrics but also to developments of online equipments for continuous monitoring of quality in textiles such as Fibre Contamination Eliminator, Intelligent Yarn Grader and Automatic Fabric Inspection.

# 5

# Fibre-growing Industry

Having gained some idea of the extent of production figures, the fibre-growing industry will be examined. As examples of typical raw materials, cotton (a plant seed-hair fibre), linen and the similar (bast) fibre jute, silk (an extruded animal fibre) and wool (an animal hair fibre) will be considered. Wool and linen almost certainly predated the other two historically, but history does not have to be followed slavishly, so the vegetable fibres, cotton and linen, will be dealt with first, followed by silk and then wool before looking finally at the artificial (man-made) fibres.

## LEAF AND STEM FIBRES

Perfectly satisfactory textile fibres can also be obtained from other parts of plants than the seed hairs, notably the stem and the leaf. Stem, or bast, fibres tend to be less ecologically harmful than seed-hair ones. Linen, a typical example, can grow with virtually no attention or fertilisers, as long as water are available. Tarres discusses the cultivation of flax in the past and predicts how it might change in the future, also describing pretreatments and textile uses. Mackie carries out a similar task with regard to hemp, another stem fibre. For best quality fibres from both sources, a relatively small amount of care needs to be taken. The climate must be very moist but mild. The selection of optimum quality fibres has traditionally been done by hand in many countries. This makes the process more costly financially, but is environmentally beneficial. In recent times, though, mechanical methods of harvesting have been introduced, with adverse effects on the environment and a reduction in the quality of the fibre produced.

Once harvested, the flax must be treated to release the usable fibres from the surrounding woody stalk and interior pith. The most important part of this process is the retting, or soaking, that allows slow decomposition of the woody parts to take place. Schulze notes the impact of flax fibre properties on spinnability. He investigates the performance of flax blended in turn with cotton, viscose, polyester and aramid components, finding that the importance of retting is paramount for successful yarn production.

Traditionally, retting was done in pools or ditches, which tend to give a better quality of product, because the advantageous moist conditions are preserved longer in their shelter. If the flax plant is cut by hand and left to ret in the ditch, then the environmental load on the planet is minimal. Jute has similar advantages; Chattopadhyay outlines the retting process and describes the physical characteristics of the fibre, mentioning its chemical composition, sensitivity to alkalis and the need to take precautions to avoid yellowing if bleaching is attempted. Sadly, modern technology cannot accept such a beneficial solution. Dew retting or tank retting is more frequently used today, using mechanical devices to turn the stalks, maintain high temperatures and remove seeds. Lennox-Kerr reports the development of a new mechanical treatment process for flax to produce linen fibres that resemble cotton in their properties and structure and can be processed on equipment designed for cotton.

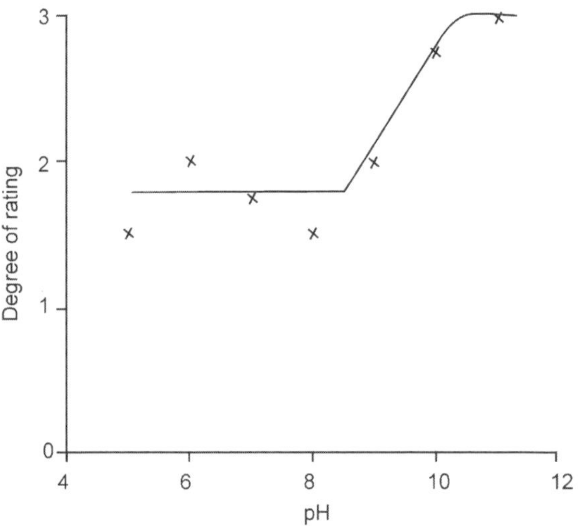


**Fig.** Effect of pH on flax retting

Mechanical harvesting speeds up the cutting process and chemical agents reduce the retting time from two or three weeks to a few days or even less. Henriksson *et al.,* for instance, propose an oxalic acid-based system for chemical retting of flax that compares favourably with the more common enzymatic retting. The authors find that retting is virtually complete if treatment is continued at 75°C for as short a time period as a mere two hours. The presence of a detergent is needed to obtain satisfactory retting. Tests of tex, tenacity and elongation properties again indicate that the fibre can be processed on cotton production equipment. Although there are continuing arguments about whether the fibre quality is enhanced or reduced by the change from manual to mechanical or chemical processing, it is the economic aspect that really plays the most important part in the debate. The end result is, once again, a damaged planet.

**Silk**

Silk production brings about a different set of considerations. The process involves rearing the silkworm grubs to the chrysalis stage, then unwinding the silk filaments from which the case enclosing this chrysalis is formed. Most commercially produced silk is of the cultivated variety, depending on feeding the worms a carefully controlled diet of mulberry leaves grown under special conditions.

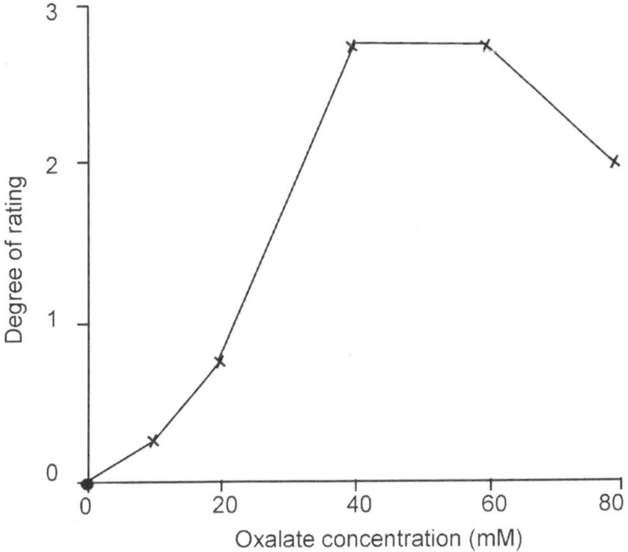


**Fig.** Effect of oxalate concentration on flax retting

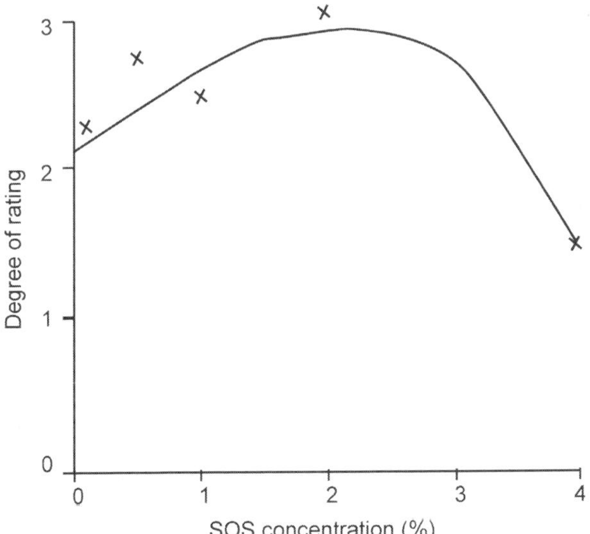


**Fig.** Effect of sodium dodecylsulphate concentration on flax retting

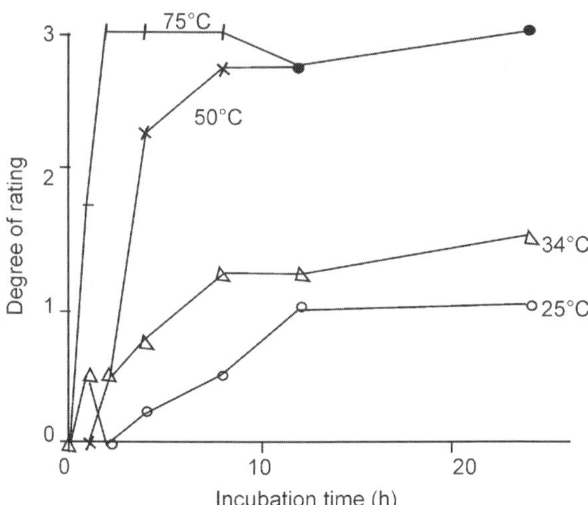


**Fig.** Effect of temperature and incubation time on flax retting

This need gives rise to the environmental costs of establishing controlled atmospheres and employing rigid growth conditions. Wild ('tussah') silk production, conversely, involves minimal interference with nature, so there is a great deal of interest in developing the fibre commercially.

Akai surveys the available techniques, focusing on the main points of manufacture and on predicting the importance of commercial manufacture of wild silk in the 21st century. Jahagirdar feels that marketing is the main constraint, in view of the lower quality (compared with cultivated silk), but stresses that the fibre can provide a major source of income for millions of tribal people in India.

Krishna Rao agree, noting that the income is year round rather than seasonal, and point out that the forest cover is an easy source of food and plants, so constituting a base for commercial development. Ghosh describes improved techniques for enhanced quality and productivity and mentions the slow transition to better working conditions and higher incomes for workers, describing advances in powered machinery for reeling and spinning. Shetty and Samson describe the production of silk without the need for traditional mulberry leaves, an obscure craft of tribal and hill peoples up to the time of writing, but suggest it as a potentially important commercial crop.

Problems shared by all types of silk are brought to light in the literature. The first of these is the need to control reeling in automatic processing, a need that Hu and Suzhou discuss in some detail. Second is the importance of understanding the effects of cocoon shape or quality on filament uniformity, a topic reviewed by Singh Maribashetty and two colleagues analyse why silkworms fail to produce satisfactory cocoons, dividing them into pathological, physiological, environmental and genetic reasons. They discuss abnormal development in the silk gland, spinneret and nerve malfunctions, and hormonal or disease causes, noting that control of both temperature and relative humidity is critical during the larval period.

Cultivated silk, in addition, needs other special conditions. Selected mulberry trees are grown to act as homes for the silkworms and their leaves are hand picked daily. The trees, like cotton plants, require fertiliser and pesticide applications, though they tend to have lesser demands than cotton plants. They also have to be selected carefully, as the silk quality is highly dependent on the diet fed to the worms. One major problem that has emerged is the need to protect them from poisoning; by agricultural fertilisers, tobacco, factory exhaust gases and biopesticides, with a list of precautions to be taken in order to minimise the effects of these modern toxins.

Thus, the carefully controlled conditions needed for silkworms to work their 'miracles' are environmentally expensive to maintain. Establishing the perfect atmosphere for the grubs needs a supply of clean air, apparatus to heat or cool to the correct conditions and power to operate the entire system. Extraction of the fibres by steaming to kill the silk chrysalis requires steam. The unwinding step depends on sensitive machines that are designed not to break the fragile threads. The cleaning process, necessary to prevent the newly wound threads from sticking to each other, uses hot water (often with detergents) to remove the silk from the gum, and the waste liquor is usually discarded to the ground water, acting as a pollutant once more. All of these components of the production process are environmentally costly.

Silk, however, is not the only animal fibre grown directly or indirectly from insects; spiders can also produce a similar filament, though their ability has not yet been adopted with any commercial success because of the difficulty of unwinding the thread without breakage. However, there is one piece of work in this general area that deserves to be mentioned as it may represent the type of future development that can be expected in the industry. Smith notes that genetic engineering has enabled a Canadian company to insert a gene from a spider into the DNA of a goat so that the goat's milk contains a spider silk protein that can be extracted and made into a filament with extremely high strength. Projected uses include medical sutures, vascular grafts, military and law enforcement protective clothing, structural engineering and packaging.

## Cotton

Cotton is grown on a bush, 0.6 to 6 metres high, in a field. Good growth of a high quality fibre requires a minimum of about 150 days of sunlight each year and plenty of fresh water. Cotton can therefore only be grown in a relatively narrow belt of the Earth's surface, within about 35 degrees latitude of the equator, and in one or two other areas of the planet where local conditions are atypical for the latitude. The quality varies greatly in different locations, with Sea Island and Egyptian cotton generally considered to be the finest and

Pakistani, Russian or Indian inferior. In order to achieve the optimum quality, efforts are usually made to enhance growth conditions. Fertilisers, insecticides and herbicides are commonly applied during the growing season. A considerable part of an international conference held in 1998 was devoted to this topic. Pertinent subjects discussed include enhancing quality by adjusting growth or soil treatment conditions and control of pests to increase yield. Daniell suggests that genetic engineering provides the best possible way of achieving both of these, since natural breeding has reached its limit because of species incompatibility and the small range of properties that can be incorporated into a plant by natural selection.

This view is echoed by Wilson, who reviews current and future biotechnology, suggesting (among less fanciful ideas) the possibility of developing blue cotton plants to avoid the need to dye cotton for jeans. Although this may appear to be a good idea at first sight, it is appropriate to wonder how environmentally costly the process will be, both in the short term, because of the technology needed, and in the long run when different shades of blue are demanded by fickle consumers or when cross-fertilisation by wind distribution has eliminated all bushes bearing natural white cotton fibres. There is also the concern, a notable one at the moment, regarding contamination by genetically modified plants of other plants, not modified by this technology, in the vicinity.

Other contributors at the 1998 Beltwide Conference, such as El-Lissy *et al.* give information about a scheme to eradicate insects, such as boll weevils or budworms, indicating that significant declines took place between 1995 and 1997. However, a conflicting note is sounded by Williams, who estimates that insect losses still represent 9.42% of the crop, the equivalent of 2.5 million bales of cotton and almost $1.5 billion dollars in financial losses. Many papers at the same conference deal with combination of biological and chemical controls to achieve sound management of insect pests, including aphids, leaf whitefly or sweet potato whitefly. Yet other contributors concentrate on soil tillage and methods of actually adding fertilisers to improve yields.

Another important aspect of cotton growth is the need for a plentiful supply of water. Drought conditions can bring about a reduction in the quality of the cotton not only because of the lack of waterborne nutrient flow but also as a consequence of changes in fertiliser effects. First, unless gravitational flow can be harnessed, water must be pumped mechanically to the site, which involves manufacturing and powering machinery for the purpose.

This machinery, like all the mechanical aids mentioned already, is itself environmentally costly. Second, the flow of water can exacerbate erosion and chemical leaching problems. Third, unless applications are carefully coordinated, the flow of water can remove some of the chemicals before they have accomplished their purpose, necessitating the application of more of them to control the growth of pests for which they are needed.

Once the plant has produced its crop and the seed hairs are open, harvesting must take place. In these days, despite the fact that hand picking can yield a better quality of product, harvesting is almost invariably done by machine, simply because the cost of labour is too high to make the relatively slight improvement in fibre quality worth the extra cost involved. This machinery adds an environmental load to the process. The timing of picking, too, is less easily controlled when mechanical picking takes place. Manual methods can allow selection of just-ripe bolls to be made, but no machine currently in existence has this capability. Another conference, this time in Bremen, included details regarding the means for carrying out standard tests for measuring maturity, along with stickiness, dust or trash levels, fibre length and high volume instrument (HVI) testing. One method suggested for estimating the time at which cotton reaches its optimum maturity uses image analysis of fibre bundle tests.

The next stage in cotton fibre production is the ginning operation, frequently taking place in the field in the immediate vicinity of the crop and designed to separate the cotton fibres from the seed fragments. Vizia and Anep claim that ginning is the weak link in cotton processing in India, because it suffers from poor efficiency and problems with cleaning and storage. They outline proposals for improvement in the future. Lugachev, in an effort to improve quality, has devised a new energy-saving technique for doffing fibres in sawtooth ginning by making use of a reflected air flow.

Anthony and Byler are more optimistic on behalf of American producers; they suggest that careful control of the ginning process (especially in the matter of moisture uptake and machine type), together with a mechanism that includes computer measurement, improves fibre quality, gives higher yield and produces fewer short fibres and neps. The scheme they recommend is one in which moisture content, colour and foreign matter are measured by standard methods, after which this information is fed forward or back, so that ginning can be increased if necessary or omitted completely if it is not needed, thus providing better separation of seed-coat fragments with minimum fibre damage. These fragments are often regarded as useless but, in a notable example of environmental responsibility, Bader *et al.* use them for cattle feed and as replacement for pine chips in poultry barns.

**Wool**

Wool is a natural animal fibre of a different kind, made from the inner fine hair of the sheep or goat. Again it is responsible for a new set of constraints. It is often advertised as the 'perfect' fibre, as a result of its desirable properties and is indeed at first glance worthy of that compliment. Though, its perfection is considerably reduced when its overall ecological impact is taken into account. Its production nowadays is complicated by a multitude of chemical treatments that add considerably to the environmental cost of achieving a fibre that is satisfactory to fussy modern consumers.

The sheep is, admittedly, a beneficial asset to the planet. Not only does it provide us with the warm fabrics that have kept out cold climates since time immemorial, but it is also a source of meat. Unfortunately, the animals that provide the best wool tend to produce poor quality meat, and vice-versa. Sheep can thrive on marginal land where virtually nothing else will grow; the moorlands of northern England where these hardy beasts make their home have only to be seen in order to appreciate just how useful they actually are in converting scrub grass into vital end products. They also provide manure to fertilise the ground on which they graze or for sale to avid gardeners.

Harvesting the wool consists of shearing the animal on a yearly or twice-yearly (depending on climate) basis. Hand shearing has virtually disappeared today, to be replaced by the use of electric shears. These do not consume a great deal of power, so are not too costly in environmental terms. But perhaps the most significant aspect of sheep growth, is the use of sheep dip, an antiseptic agent, to prevent infection or to remove insects and other pests from the animal's coat, which (depending on the exact formulation used) may seriously contaminate water supplies.

## PREPARATION OF SURFACES FOR FABRIC COVER

Often a builder or aircraft restorer will have the mistaken idea that fabric covering begins when you cement the fabric in place. This is far from reality. As a matter of fact, a lot of time and effort will be needed prior to ever cutting the fabric for placement on the aircraft. Anyone who has ever restored an airplane certainly knows that most of the total work involved is in the preparation phase. A few basics need to be presented concerning adequate preparation. First of all, if you are recovering your airplane take care in removing the old fabric. You can save yourself a lot of time and effort by carefully cutting the old fabric away leaving the inspection plates, drain grommets, reinforcement patches, and control cable cutouts in tact.

This will help you with placement of these items on the new fabric. Secondly, always make sure you use epoxy primer on metal and fiberglass parts and epoxy varnish on wood parts. Why—because the majority of other primers and varnishes will be lifted from the aircraft surface by the chemicals found in covering processes. MEK and reducers found in covering process chemicals will not affect epoxy primers and varnishes. They will lift other paints and varnishes like a paint stripper allowing moisture to collect in the metal or wood with obvious consequences. If you have already primed or varnished with some other product test it before applying the fabric. Soak a rag with MEK and leave it on the surface for about 30 minutes. If it lifts the paint or varnish you need to redo the surface. Often you can simply spray epoxies over the existing finish without having to strip them. This too, can be tested. Spray the epoxy over the existing surface on a small area to be sure it does not act as a paint stripper prior to applying it to the entire piece.

Before priming or varnishing a surface be sure it is completely clean and free of all oil and other contaminants. Do not prime over rusted pieces. Remove the rust and immediately prime. A piece of bare steel will rust within hours if a primer is not in place. Make sure you fill all dents in leading edges, etc. I would recommend a product called SuperFil rather than Bondo. Bondo is a polyester filler that will shrink with age. I do not recommend using Bondo on aircraft surfaces unless you are prepared to redo the filled portion of the surface after it shrinks and cracks the topcoat. SuperFil is an epoxy filler. That means it will not shrink over time. It may be used on wood, fiberglass, or metal with equally good results. You may want to use a cloth padding on a really dented leading edge. Polyester padding is often used between the leading edge aluminum or plywood and the fabric itself.

All of the sharp edges that could potentially cut the fabric should be covered with anti-chafe tape. This usually involves rivet heads, metal seams, and sharp edges. Let your sense of touch be your guide. If you feel something sharp cover it with the anti-chafe tape. Do not use masking tape for this purpose. It will retain moisture and cause problems later on. Also, paper masking tape will turn brown with age and possibly show through a light coloured paint.

On your wings you will want to ensure the ribs are parallel to each other and aligned properly prior to placement of the fabric. Inter-rib bracing tape is used for this purpose. It will keep the ribs straight up and down when the fabric is heat tightened over them. When complete, the inter-rib brace looks like a series of "Xs" in each rib bay.

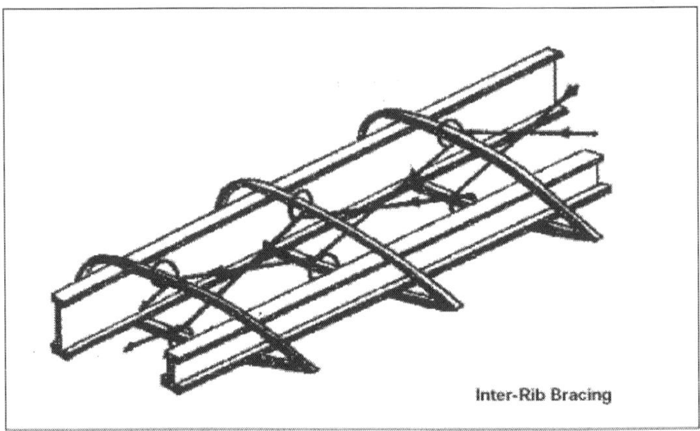


Fig. Selecting Proper Fabric Weight

This decision has a major impact upon the life of the fabric on your aircraft. If you use a fabric too light in weight for your airplane you will certainly have associated problems. First of all, lets look at what fabrics are available. Grade A cotton was the fabric of choice until the 1960's. It was replaced with polyester fabrics designed to be shrunk with heat. To my knowledge, Grade A fabric is virtually unavailable today. If you do find it make sure it meets the FAA standards required for production aircraft and test it prior to placing it on your airplane. You could get old fabric that has been exposed to sunlight and has lost its strength before you even put it on the airplane. My advice is to stay away from Grade A cotton today. Polyester fabric is much easier to apply.

Polyester fabric comes in different weights and strengths. The most common fabrics are light weight, medium weight, and heavy-duty. Light weight fabric weighs 1.7 ounces per square yard. That equates to a total fabric weight of about 9 pounds on a J-3 Cub size airplane. Medium weight fabric weighs 2.7 ounces per square yard bringing the total weight of fabric only for our J-3 to 14 pounds. Heavy-duty fabric weighs in at 3.4 ounces per square yard bringing our total comparative weight up to about 18 pounds for our J-3. Typically, the coatings will add 30-40 additional pounds to the overall weight. This is dependent upon the process applied. The strength of fabrics varies from a breaking strength of about 70 pounds per square inch up to about 130 pounds per square inch, light weight versus heavy-duty. Selection of fabric for your airplane is simple. If you are covering an ultralight, glider, or an airplane with a small horsepower engine (below 65 HP), you can safely use light weight fabric. If you are covering an aerobatic aircraft, large bi-plane, warbird or crop duster, in short any high wing loading aircraft operating under severe conditions, use heavy-duty fabric. All other airplanes (this constitutes the majority) should use the medium weight fabric. You cannot go wrong using a heavier weight fabric. Problems are much more likely to occur when you use a lighter weight fabric than the one recommended for your airplane. These problems surface in the form of damaged coatings on top of the fabric caused by movement of the fabric itself in the form of drumming or beating. So, if in doubt, add a few pounds and go with a heavier fabric.

By the way, Ceconite, Superflite, and Poly-Fiber fabrics are all basically the same. They are polyester fabrics woven at the same mill on the same looms. They are all capable of being shrunk about 10% and the thread count is the same on the warp and fill. That means you do not have to worry about the orientation of the fabric. It can be placed on the surface in any direction.

## FOLLOW THE INSTRUCTIONS

This is a key point. You must follow the instructions written by the fabric process manufacturer. Reviewing the STC's mentioned earlier, on a production airplane you cannot mix fabric covering processes. Not only is this a legal issue regarding the airworthiness of your airplane but it also can result in having to recover your airplane prematurely. As an example, Poly-Fiber coatings are not compatible with nitrate and butyrate dopes. If you put Poly-Brush (the first chemical coat on the Poly-Fiber process) on fabric as your first coat of chemical and then switch to nitrate dope for subsequent coats you will have problems in the form of cracking and peeling. These problems usually do not appear right away but they wait until you finish the job and fly for a few months.

Then they appear. What is the solution—recover the airplane. By the way, if you do have major fabric problems such as cracking and peeling of coatings, it is impractical to strip the coatings off the fabric. The best course of action is to recover the airplane. Even better, learn from other people's mistakes and follow the instructions and you will not have to recover for 15+ years.

Another area of concern is experimenting with a proven system. Listening to so-called "experts" who provide you with a quick and easy way to short-circuit the covering process. I have encountered individuals who are experimenting with various steps of the covering process from the type of fabric cement they use to using latex house paint as a final topcoat. Let me say this very clearly, the fabric coating manufacturers have tested their products. To my knowledge none of the tests involved latex house paints. In other words, *if it is not in the manual don't do it* unless you want to spend another $2,500 to recover the plane.

## FABRIC TAUTNESS

How tight should the fabric be before you start putting on the coatings? This is a very good question and an additional problem area. If the fabric tautness is not correct you may have difficulties. If the fabric is too tight you may do damage to the structure itself. If too loose, you may have the coatings crack and peel due to too much movement. In any case, take the time to properly shrink the fabric. Remember, the fabric is capable of shrinking about 10%. Keep that in mind when you cement the fabric in place. You should initially put it on the surface so that it is snug—a technical term meaning neither too loose nor too tight. A good rule of thumb—on a surface such as a wing you should be able to lift the fabric about 1 inch above a rib prior to shrinking.

One thing is certain; you must use an iron to shrink fabric. If you have a heat gun hide it until you have covered the airplane. You cannot control the temperature emitted from a heat gun. It is absolutely essential that you shrink the fabric using an accurate temperature from the iron. This can only be achieved using an iron that is rated at 1100 watts or higher. You must calibrate this iron for accuracy. When you shrink the fabric it must be done at a precise temperature within + or – 10 degrees. The initial shrinking of the fabric is done at 250 degrees F. At this temperature the fabric will shrink about 5%. If you are using nitrate and butyrate dopes this is the final temperature that you will use. Nitrate and butyrate dope (yes, even non-tautening) will continue to shrink fabric over a long period of time. If you have an ultralight or smaller aircraft you may want to stop shrinking at 250 because of structural considerations. Fabric, when being shrunk, has incredible power to bend and twist. You should watch for evidence of this while using the iron.

Most aircraft structures require a final shrinking of 350 degrees. This ensures proper tautness of the fabric. Do not exceed 350 degrees. At 375 degrees the filaments of the fabric begin to thermo-soften and actually loosen on the structure. If this occurs the fabric will appear as though it has not been tightened enough. At 415 degrees the filaments begin to melt. For this reason it is absolutely imperative that you calibrate your iron using a glass thermometer. I do not recommend using the spring type thermometer. It is not as accurate as a bulb type. The Poly-Fiber manual has a very good explanation as to how to calibrate an iron.

## PERCEPTIONS OF FABRIC

In recent years there has been a dynamic shift in attitudes toward textiles, and a new type of textile designer has challenged traditional forms of construction and content. As a result of a broadening vocabulary of materials, processes and techniques, textile designers are faced with more creative choice than ever before. New sensitivities and responses are also called for. On another level, attitudes and activities within the practice now mirror all facets of contemporary life; textile designers can be radical, funky, technical, conservative or serene; tradition has loosened. Many of the exciting changes that have happened in fabric in recent years have acted as a creative stimulus for other design disciplines.

This versatile medium now supplies the worlds of fashion and interiors with an increasing and sometimes challenging range of materials and products. The fashion designer now constantly looks for original, innovative materials to work with, making fabric a fundamental part of contemporary fashion statements. For the interior designer or architect, new textiles offer everything from the complete co-ordinated chilled look to eclectic zany mixes. Whether they are looking for harmony, drama or functionality there is always one or more of these on offer, and quite often all three of them at the same time.

An increasingly diverse and sophisticated consumer market has witnessed innovations in fabric blends, the use of new technologies and the amazing march of synthetics. The effect of this combination has been the ability to explore new surface qualities of matt and shine as well as new combinations of weight and performance, and to reinvigorate old yarns and traditional fabric patterns. The look, the feel and the drape of fabric has entered a new era. No longer is the use of materials restricted to a perennial function: sportswear enters street wear, safety and protective clothing can influence couture. These dynamic fusions in their own way stimulate further opportunities - there are new surfaces to print on, and fabric development starts to play with surface qualities as variable as the difference between paper and rubber, between moss and cellophane.

**New Fabrics, New Aesthetics**

The fusion of technology and tradition has revealed itself in a diverse assortment of new fabrics and materials, impeccably mirroring changes in lifestyle and the way we embrace modernity and progress in the twenty-first century. Increasingly the gap between the producer of textiles and the consumer is shrinking, enforcing the demand for innovation. The merger of tradition and technology applies to almost every aspect of the textiles world; from designers to craftspeople to artists and industry.

Lifestyle culture has also been responsible for blurring the relationship between these disciplines and creating new intermediate markets and opportunities. Similar changes in the world of fashion create even more hybrid designers and virtuoso applications of fabric. Consequently, during the last decade of the twentieth century, a new breed of designers came to the forefront of textiles, and their work sits as comfortably in a gallery as it does on the shelf in Barneys, Liberty, Bergdorf Goodman or Harvey Nichols. The 'new' textile professional has foresight and employs intuition, working in a space where there are no limitations or restrictions.

Advances in textile design have had a remarkable effect on fashion. In addition to exploring issues of silhouette, the fashion designer envisages material innovation as an important motivating factor in his or her creative process. On occasion the material itself may well inspire silhouette. The creative collaboration between the fashion designer Issey Miyake and Makiko Minagawa, the textile director of Miyake Design Studio, illustrates this. Today's textile designer is resourceful and multi-talented, and can be found either in a laboratory breaking new ground through developing hi-tech materials, or leading the way by broadening the vocabulary of advanced decorative effects.

This is often achieved by means of investigating processes and treatments such as moulding, embossing, heat-bonding and sculpting as practiced by Sophie Roet and Nigel Atkinson. Material innovation in textiles involves surface exploration which can now result in novel, sometimes curious and extraordinary fabrics. These are attained by way of combining various methods of production and finishing. Outcomes can be exciting but poised carefully between the worlds of commercialism and discovery.

More unconventional approaches to making and designing are often stimulated by developments in yarn and fibre technology. These constantly enhance the creativity of the textile designer, as do innovations in textile processes which direct us away from traditional notions about the subject. Yarn innovation has led to avant-garde mixtures of fibres, including the silk and aluminium 'Inox' yarn and the wool and stainless steel blend from 'Bekintex'. These blends not only boost the properties of natural fibres but also usually offer additional qualities such as reducing static. Besides possessing performance-enhancing benefits, yarns are also developed for aesthetic reasons in order to achieve specific visual or textural effects. Synthetic fibres are often treated to mimic the intrinsic qualities of natural fibres or to create unusual, quirky finishes such as crunchy, glassy, clear nylon or novelty rubberized wool that feels like paper and looks like bark.

The recent technical and aesthetic progress in textiles deserves wider recognition. Today, the subject includes remarkable fabrics, distinctive furnishings, fascinating artworks and spectacular fashion. We are witnessing an era that is historically unmatched in terms of invention and experimentation in the field of textiles. An increasing number of designers are revealing the scope of the subject and the versatility of textiles as a creative medium. Through its major trade fairs, scattered across every continent, textiles now manages to disseminate these innovations rapidly across the globe.

**The Evolution of Textile Practice**

Conventionally, textile designers will have a strong grounding in one or perhaps two of the traditional specialisms of woven, printed, knitted and embroidered textiles, and these are still considered fundamental skills in textile practice. However, now is perhaps the right time to question and dispute these traditional categorizations of fabric. After all, current material experimentation within the discipline takes the subject beyond the realm of the yarn and the cloth. Technological advances, sophisticated consumerism, changes in society and lifestyles, diverse innovative thinking and media and communication touch not only textiles practice and philosophy but also the wider sphere of design disciplines in general.

Textiles is closely implicated in disciplines as diverse as fashion, visual communications, product design and interior design. Attempting to maintain the separate quality of traditional textile practices has proved difficult: as the subject has evolved, there has often been contention over the value and purpose of maintaining the purity of techniques and approaches; calls for change, and calls against it, immediately face an onslaught of criticisms. Textiles is now obviously something greater than the craft disciplines it has grown from.

Given this, these elementary skills still remain the foundational experience of textile practice, and they provide the textile professional with an initial framework. However, what follows the acquisition of those first skills is rather more complicated to describe or constrain. Simply describing textiles as printed, constructed or embroidered is not adequate. Crossovers occur continually, and they are not only within the disciplines mentioned here: textile designers dabble in history, culture, industry and science. The language of textiles has thus expanded through the absorption of technical knowledge or artistic ambition drawn from other subjects. With inventiveness and vision, new dimensions have been applied to time-honoured skills and materials and a case can now be made for the redefinition of textile practice.

Isabel Dodd is one of the new generation of designers whose alternative approach to textile design securely positions her in the future. Working with the most modern of synthetic materials, her printing process causes the surface of stretch jersey and velour to crinkle and pucker, producing marks similar to those formed through embroidery. We could loosely categorize her as a printer, simply because she uses a print screen. However, she is actually a typical example of individual textile designers who transcend their original formal training in order to interpret and communicate new creative ideas.

We may think that jumping from one specialist practice to another is daunting, yet many contemporary textile designers use whichever methodology is most appropriate for them to realize their conceptual ideas. In many ways this method of working is far from being a hindrance - in fact, a greater knowledge of process, by and large, enhances creative aptitude. Dodd's fabrics have been described as unusual, curious, avant-garde and sculptural, but whichever way they are explained, they are not what we traditionally associate with the conventional tools she uses.

**The Influence of Skill**

Fusions of traditional skills and new technology question whether textiles practice sits more comfortably in the scientific and industrial arena or whether it is essentially hands-on and skills-based. The introduction of new technology raises the question as to whether the practice could be branded as either purely scientific or, more traditionally, as an expression of an aptitude with material. The new concerns within textile practice have encouraged a fresh approach to a traditional discipline; one that is now in transition and appeals for re-evaluation as amply demonstrated by the recent 'A Twist in the Yarn' conference organized by Alan Holmes at Manchester Metropolitan University.

There is currently much research taking place in the textile world, exploring technologies related to textile production, and design issues involving the relationship between surface and structure. There are also a number of textile professionals involved in material development and research who would challenge any predetermined views of what textiles constitutes. Sarah Taylor, Janet Stoyel and Janet Emmanuel are among those who are revolutionizing our perceptions of fabric.

Sarah Taylor has become somewhat of a pioneer in the field of weaving, having worked in collaboration with the Japanese company Mitsubishi Rayon. Her research centres on developing light-emitting fabrics through the use of fibre optics. The close supportive relationship she has with industry is vital in the development of her technologically advanced woven fabrics. Despite the craft origins of Taylor's work, perhaps industry on more than one level has historical sympathies with her practice - after all, the implicit data at the heart of weaving is easily convertible for mass production in industry:

There is a fluidity in the practice, design and art of woven textiles that enables textiles to fit easily with contemporary technology. A textile maker or designer who works at a small craft-shop level producing one-off pieces can, from the same conceptual base and using the same equipment, produce samples that industry can convert without any fuss for factory production.

Fibre optics have been used in protective clothing for military use, but Taylor has been investigating their possible applications within interiors, perhaps as multifunctional mood-enhancing design pieces. Through controlling the colour of the light released from the fibres, Taylor proposes that it would be possible to agitate or calm the environment. Cheryl Adnitt is also developing methods of incorporating light within her fabrics.

Her approach differs from Taylor's in that she uses either translucent monofilament which glows when exposed to ultraviolet light, or weaves conductive wire and LEDs into her fabrics, which illuminate when connected to a power source. The latter part of the twentieth century witnessed a definite move in the direction of a rather more scientific, technological and laboratory-based approach to textiles. This combination of chemistry, physics, textiles and design has conjured up numerous ground-breaking fabrics.

Janet Stoyel's work fits into this new category of practice, and she would no doubt agree that industrial liaison is advantageous and sometimes essential when working on developing hi-tech processes through innovative design. Stoyel is recognized as leading the way in the use of lasers for producing textile designs.

Conventionally, ultrasonic technology has been employed for various purposes in the medical and industrial worlds. On the whole, within the textile industry, it is a relatively new discovery, first appearing on the scene in the early 1970s. For Stoyel to practice her design skills, it was crucial that she should be able to demonstrate proficiencies in areas other than textiles. There was a significant learning curve before she was able to operate the sophisticated machinery she needed in order to create the sculptural pieces she envisaged. Stoyel's range of skills can be compared to that of other textile designers such as Dodd and Taylor, already mentioned.

Much of Stoyel's material explorations conform to her unorthodox rules in that she blends synthetic metal and polymer textiles with such natural materials as silk and paper. The UK Ministry of Defence assisted Stoyel in her efforts to unify textile design and technology. Through utilizing their expertise she was able to establish her process of cutting using lasers and welding with ultrasound in order to create three-dimensional constructions.

Following in Stoyel's footsteps is Janet Emmanuel; although they share common ground in the equipment they use, their reasoning for making and the culmination of their efforts is quite different. Emmanuel's initial interest in technology can be narrowed down to a period at the end of the twentieth century when creative professionals from all design disciplines began to exploit new methodologies and materials.

Her preliminary studies within the subject focused on origami techniques, initially in silk organza and then in synthetics: these evolved into pleating ideas which were explored through traditional and industrial means, the goal being for the forms to become permanent. Some may feel her personal inspiration is somewhat obscure, but the driving force behind her research was to explore sound. In contrast to twentieth-century artists such as Kandinsky and Yves Klein who were fascinated with the visualization of sound (interpreting the relationship between colour, music and mark), Emmanuel opted to translate and examine what she calls silent sound, manifesting this through material manipulation.

**The Influence of Science and Engineering**

It is very difficult to escape the fact that science and engineering play a huge part in textile industries; much of the textile 'engineering' that occurs is usually intended for the manufacturing sector, but can also greatly influence design. An observation of history reveals that there have been major technical advances which in time have proven to be revolutionary for designers. Among these have been man-made regenerated fibres, synthetics and more recently smart or intelligent fibres.

One reason behind the original development of materials that mimicked natural qualities was to provide the consumer with an alternative to natural fibre at a fraction of the cost. The purpose for developing these materials, which would act as a substitute for the authentic article, was at some point superseded by the idea that we developed fabric just because we could. The scientists and engineers were 'able to' enhance, improve on and surpass the previous advances, so they did. Issues like fabric performance were stimulated by specialist sectors such as the sportswear industry; coupled with increasingly sophisticated consumer requirements, conditions were perfect for developments to continue.

The first 'engineered' fibres were generated from organic material and were developed to mimic the qualities of natural yarns, enabling the production of artificial silk and wool fibres. Developed in the second half of the nineteenth century, these cellulose-based fibres, constituted from wood pulp, were chemically altered so as to compose yarn. Following this development many years later, under the influence of environmental pressures, a 'green' recyclable artificial fibre was launched, again based on wood pulp. Apart from the introduction of microfibres (very thin, fine versions of synthetic fibres), Tencel is the first new man-made fibre to have been introduced since Polyolefin in the 1960s. It is renowned in the industry for being the 'chameleon' of fibres as it is able to completely imitate the qualities of other fibres, at the same time enhancing their strength.

Synthetics on the other hand were developed in science labs from such substances as coal, oil, other minerals, glass and carbon, after the discovery of regenerated fibres. There are two prominent established players in the world of synthetics, which have now become household names, Du Pont and ICI. Much of the initial research underlying the development of synthetic materials took place in the 1930s, while manufacturing became firmly established in the 1950s.

Many synthetic fibres such as nylon, acrylic and DuPont's famous brand Lycra are household names. Polyester is now combined with its many natural counterparts, and is the most commonly used synthetic fibre, recyclable from clear plastic bottles. Synthetics have become infinitely sophisticated since their introduction microfibres and elastanes have been developed which possess virtues of highperformance and handle that natural fibres would find difficult to match.

In comparison to the innovations already mentioned, smart fabrics are in a league of their own. The sportswear industries drive the textile researchers, designers and technologists to create increasingly sophisticated materials to meet their demands. On first impression the needs and desires of the end user in this sphere could quite easily be confused with the stuff of a science-fiction novel. Materials that react to the body's needs are of the utmost importance in the sports world.

More often than not, these materials are developed to enhance performance as in the case of the Sydney 2000 Olympic gold medal winner, Ian 'Thorpedo' Thorpe, who hit international headlines when he wore the Adidas hydro-dynamically styled 'Fastskin' swimsuit. This concept was actually originated by the Britain-based Speedo swimwear company to reduce drag in the water. They state that the conceptual basis for their material was sharkskin, and that 'Fastskin' was designed to imitate the way in which a shark's scales enable it to glide smoothly through the water. The theory of copying naturally occurring phenomena like this is not new and, as engineering skills increase, such copying is likely to happen more often.

An increasing trend is that fabrics and yarns developed for a specific application or industry are used by other product sectors. For example, fabric technologies originally intended for industrial use or medicine can now form a central part of sport or fashion. Both of these worlds are colossal industries in their own rights and it is difficult to ignore the volume of the textile components used within each of them. In order to indicate the scale of the industries we are examining, a recent study by the American Georgia Institute of Technology, revealed that sport alone is a bigger industry than a combination of the television, radio, motion picture and educational service industries. Dr Isa Hoffman, Project Manager at Interstoff, Frankfurt, Germany, sums up current opinion on smart materials:

It is the fabric that makes the trend. Garment physiology and environmental behaviour, comfort and lifestyle have become the components of modern fabric creations and turned them into high-fashion, high-tech products ... Words such as 'intelligent' or 'electrotextiles', 'anti-bacterial' or 'biomimetic' could well come from a computer dictionary. Today's fabrics are 'programmable'.

In beginning to reappraise the current condition of the textile world, we note that the 'future model' of the material supplier may also be transformed. As the definitions within textile practice begin overflowing into the worlds of science, engineering and technology, the fashion designer, product designer or architect that seeks innovative, original and new materials will inevitably search for a source other than that of the conventional fabric fair.

One such 'future model' of the fabric and materials supplier is George M. Beylerian, founder of Material ConneXion. He offers the facility to track down the most advanced materials and their suppliers from all over the world through his resource exchange. Additionally, he provides the expertise to assist with market research, design sourcing and product development; in doing so, he connects the design community with its technological counterpart.

Within this model, Beylerian is keen to work with material craftspersons, believing their role to be that of creative philosophers, and using their various skills and methodologies to manipulate and control new materials. It is envisaged that the exploration and assessment of the intrinsic properties of these materials will reveal the scope and manner of their application. This knowledge would lead to further material evolution. The role of the material supplier, as Beylerian himself puts it, may well be as 'an intelligence bank for the future'.

**Re-evaluation**

Definitions within contemporary textile practice are now open to debate, and traditional boundaries between sub-disciplines have become blurred. No longer does a weaver just sit at a loom, or an embroiderer at a sewing machine. Today, such persons, while engaged in their creative process, are to be found in front of computer screens, in science labs or engineering departments, or working alongside specialists from other disciplines. The creative processes now found in textiles defy the historical divisions and definitions applied to it. Currently, many practitioners who consider themselves textile specialists are occupied with roles and tasks not conventionally linked to the textile world. Textiles is full of contradictions which actually contribute to the innate nature of the subject, and simultaneously slants towards craft and towards technology.

Complications and confusions can also arise in its literature: we find printed textiles and felt-making featured in embroidery journals, craft periodicals don't have a problem with presenting textile art, and pieces on embroidery are included in felt-making publications.

Since much textiles practice fuses invention, design, craft and technical skill, each discipline within the textile arena seems to fit snugly into another. This fusion is not some abstract theoretical hope - it is a self-evident reality, populated by real textile people. Among the many designers that fit into this rather blurred new categorization of '1/1/textile professional' is Kei Ito. She investigates materials, amalgamating organic structures with ultramodern fabrics, manipulating them to form clever, complicated outcomes. Ambiguities arise when an attempt is made to categorize the work, which sits easily between textiles, craft and fashion art.

With the new approach to textile making come new approaches to evaluating the practice and the medium. Freddie Robbins attempts to intellectualize the physical realm of knitting and compares it to current technology, making observations about our perceptions

There are many similarities between knitting and the internet. They are both physically solitary processes.

Both are making sense of, translating and communicating information. Computers and knitting patterns use grids, with coloured squares to relay visual information. With both mediums it is easy to 'undo' or 'unravel' your work. However, knitting is an 'everyday skill' of the future which crosses social, cultural and geographic boundaries.

The reappraisal of textiles has repercussions in education and in the definition of the textile industries in general. Over the last few decades of the twentieth century textiles gradually reformulated itself, perhaps not intentionally, but nevertheless it is now not what it used to be. This evolution is attributable to a new generation of inquisitive designers and the ready availability of exciting, versatile and innovative raw materials.

All this makes the terms print, weave, knit and embroidery suddenly quite archaic, and one could almost say obsolete, as the divisions between them are eradicated. We are truly at a point in time where re-evaluation is needed; this is even reflected in the nature of the student experience within the subject, many no longer desiring pure and simple textile design. There is a craving for something altogether more definitive of the twenty-first century, textiles of the future. As a result of the perpetual changes in textiles technology and culture the impetus to redefine textiles will undoubtedly grow.

## FIBRE DYNAMICS IN THE REVOLVING-FLATS CARD

Over the last 30 years numerous developments have taken place with the cotton card. The production rate has risen by a factor of 5 with the main rotating components running at significantly higher speeds. Triple taker-in rollers and modified feed systems are in use, additional carding segments are fitted for more effective fibre opening, and improved wire clothing profiles have been developed for a better carding action. Advances in electronics have provided much improved monitoring and process control. Most of these developments have resulted in enhanced cleaning of cotton fibres, reduced neppiness of the card web and better sliver uniformity.

Despite the various improvements made to the card a commonly held view is that more is known about the cleaning processes on the card than about the carding process itself. For instance, modern cards can achieve an overall cleaning efficiency of 95%. It is well established that the cleaning efficiency of modern taker-in systems is a round 30%, that the cylinder/flats action with the latest wire clothing profiles gives 90% cleaning efficiency and that effective cleaning is associated with lower neps in the card web. However, even though the nep content and the sliver Uster CV% are used as quality measures of carding performance they are not satisfactory indicators for anticipating yarn quality. This is because some fibre arrangements in the sliver may lead to nep formation and imperfections during up-stream drafting processes. In addition to the removal of trash and neps, important aspects of the carding process in relation to yarn quality and spinning performance are the degree of fibre individualisation, the fibre extent and the fibre hook configurations in the sliver. With regard to these factors, increased production rate can reduce carding quality. It is therefore of importance that a better understanding is established of the effect that carding actions have on such quality parameters, particularly at high production rates.

The most widely accepted view of how fibres are distributed within the card under steady-state conditions. Reported studies into the fundamentals of the carding process have largely been concerned with how the principal working components of the card affect this distribution of fibre mass and interact with the mass to achieve: trash and nep removal from cottons; the disentangling of the fibre mass into individual fibres, with minimal fibre breakage; and the alignment of the fibres to give a sliver suitable for drafting in down stream processes. These actions occur at the interface of the card components within the three zones. This paper therefore gives a critical review of published research on the:

- Mechanisms by which the fibre mass is broken down into individual fibres,
- Mechanisms of fibre transfer between the component parts of the card · effect of the saw-tooth wire geometry on these actions

**Zone 1: Fibre-Opening**

Separation and Cleaning of the Input Fibre Mass: The taker-in has effectively a combing action, which results in the breakdown of the tufts, consituting the fed fibre mass, into single fibres and smaller size tufts (tuflets), and in the liberation of trash particles ejected from the mass flow by the mote knives positioned below the taker-in. To effectively breakdown the fibre mass feed into tuftlets with minimal fibre breakage, the taker-in wire has to be coarse, with a low number of points per unit area (4.2 to 6.2 $pcm^{-2}$) and not too acute an angle of rake. The objective is to obtain gentle opening of the fibre mass feed and easy transfer of the tuftlets to the cylinder. Angles of 80° - 85° are used for short and medium length cottons to give effective opening and cleaning. For longer cottons and synthetics, a 90° or negative rake may be needed to facilitate gentler opening and satisfactory fibre transfer to prevent lapping of the taker-in.

Fibres, usually very short fibres, which are not adequately held by the teeth or present in the interspaces of the clothing are ejected causing fibre loss. However, it is the mote knives that govern the amount of fibre to trash (*i.e.* lint) in the extracted waste. Experimenting with the settings of two mote knives below the taker-in, Hodgson] found that the absence of the knives greatly increased the lint content with little increase in trash. With the knives present, the best setting was that which gave the least waste since increasing the amount of waste did not improve cleaning. Artzt found that irrespective of teeth density and tooth angle the waste increased with taker-in speed but the increase was attributed to higher lint content.

It is reasonable to assume that the smaller the tuftlet size and the greater the mass ratio of individual fibres to tuftlets the better the cleaning effect of the taker-in. Supanekar and Nerurkar suggest that the taker-in breaks down the fibre feed into tuftlets of various sizes and mass, conforming to a normal frequency distribution. In the case of cotton, some tuftlets may consist of only fibres whilst others will contain seed or trash particles embedded among the fibres, these tuftlets constituting the heavier end of the distribution curve. Thus, the mean of the distribution would depend on the trash content of the material, as well as on the production rate, the taker-in speed and the wire clothing specification.

However, the authors did not report any data to support their ideas. Little detailed information has yet been published on the mass variation of tuftlets or on the relative proportion of discrete fibres to tuftlets resulting from the combing action of the taker-in. Nittsu using photographic techniques studied the effect of process variables on tuftlet size. It was found that the total number of tuftlets decreases the closer the feed plate setting, the lower the feed rate, the smaller the steeper rake of the saw-tooth clothing and the higher the licker-in speed. Since the licker-in opens the batt into both tuftlets and individual fibres, a decrease in the total number of tuftlets suggests an increase in the mass of individual fibres. Liefeld calculated estimates of the opened fibre mass at various stages through the blowroom and gives a value of 50mg for tuftlets on the taker-in. Mills claims that the calculated optimum number of fibre per tooth is one, and that this should be maintained at increased production rates by increasing the taker-in speed. There is, however, the question of fibre damage at high taker-in speeds.

Honold and Brown found no fibre damage occurred at speeds of up to 600 r/min. Krylov reports the absence of fibre breakage at speeds up to 1,380 r/min, and Artzt's work shows taker-in speeds to have a negligible effect on fibre shortening and subsequently on yarn strength. In all cases cotton fibres of 26.5-30.2 mm (2.5% span length) and 3.8 - 4.9 micronaire were processed. The level of fibre breakage, however, would seem to depend on production rate and the batt fringe setting to the licker-in. High production rates achieved by increased sliver counts and a close setting of the batt fringe result in significant fibre breakage. No fundamental studies have been reported on the forces involved in the fibre-wire interaction of revolvingflat card components. However, Li and report a simulated study of fibre-withdrawal forces for wool in high-speed roller-clearer cards. Although impact forces could cause damage, it was found that card component speeds had no significant effect on the withdrawal-force, and that fibre configuration and entanglement were the important factors.

The importance of producing small size tuftlets is evident form the various components fitted in the fibre-opening zone on modern short-staple cards. Saw-tooth wire covered plates, termed combing segments, fitted below the taker-in or built into the taker-in screen are claimed to give improved trash removal.

Reportedly, the stationary flats fitted between the taker-in and the revolving flats provide extra opening of the tuftlets transferred to the cylinder from the taker-in. They also act as a barrier to large, hard, trash particles such as seed coats, protecting the wire of the revolving flats from damage, particularly at high cylinder speeds. This enables finer wire to be used for the revolving flats and thereby improves the cleaning effect of the interaction between cylinder and revolving flats. The chances are also reduced of longer length fibres becoming deeply embedded in the revolving flats to become part of the flat strips. These attachments are widely accepted by the industry as beneficial, particularly at high production speed. However, there is no published systematic study of their effectiveness in reducing tuft size, and the effect of stationary flats on the recycling layer, $Q_{2,}$ is unknown. The little information that is available attempts to illustrate the effectiveness of these components on yarn quality, but there is no evidence of analytical rigour in the way the data were obtained.

The effect of the combing segment and the stationary flats on dust deposits in rotor spinning and on the imperfections in several types of ring spun yarn. The figure includes values for the effect of stationary flats above the doffer. It would appear that the added components in the taker-in region might well reduce the dust deposit in the rotor, but the results showing improvements in yarn quality are not convincing, and in all cases the stationary flats above the doffer appear the most effective.

Leifeld reports that the cylinder - revolving flats carding action occurs when the fibre mass delivered to the cylinder is in a highly opened state. Tandem cards are said to give a high standard of carding with low nep and trash levels in the card web. This is because a uniform web of almost discrete fibres is fed to the second cylinder of the tandem card and closer revolving flat settings with higher cylinder speeds can be used. Single taker-in systems, even with combing segments and stationary flats, cannot give as high a degree of opening. However, Leifeld reports that a triple taker-in system facilitates high taker-in speeds and, fitted to a single-cylinder card, feeds a uniform web of discrete fibres to the cylinder, thereby offering a more cost-effective process than the tandem card, but no comparative data for the two types of card are given. Although it may be reasoned that a triple taker-in action should improve nep removal, it is of importance to compare the web qualities with regard to dust and trash content, the level and type of fibre hooks, and the degree of fibre parallelism since these greatly influence yarn quality.

Contradicting the triple taker-in approach, Mills states that the fibrous material fed to the card should not be broken down into individual fibres by the taker-in system. This is because the fibres would remain largely disoriented with a high proportion of them lying transversely to the direction of mass flow when transferred to the cylinder and subsequently to the revolving flats. This can result in fibre loss during transfer to the cylinder and an unevenness of the fibre mass across the cylinder width, causing neps to be formed and degrading the carding action between the cylinder and the revolving flats. It is claimed that good carding requires a thin, uniformly distributed sheet of well-opened tuftlets fed to the cylinder from the taker-in. Fujino reports results that would appear to confirm the view that as the level of opening increases through faster taker-in speed, the degree of fibre parallelism on transfer to the cylinder decreases. The nep level in the card web was, however, observed to decrease noticeably with increased taker-in speed. This was attributed to the reduced speed ratio of the cylinder and taker-in. Artzt found that reducing the taker-in/cylinder draft ratio from 2.4 to 1.4 caused yarn imperfections to increase. In contrast to these findings Harrison states that increasing taker-in speed did not affect the nep level in the card web, the exception being for low micronaire cottons. The apparent contradictions in these results suggest that a better understanding of the transfer mechanism may be needed which takes into account fibre properties.

**Fibre Mass Transfer to Cylinder**

Two contrasting views have been reported on the mechanism of fibre transfer. Oxley suggests that the fibre mass on the taker-in is ejected between the cylinder wire and the back plate. Whereas Varga believes that the fibre mass is stripped from the taker-in in the following way. In the feed to the card, tufts and fibres lie randomly and by the action of the taker-in are brought into length-wise orientation in the direction of the roller rotation. The trailing ends of newly formed tuftlets protrude above the taker-in wire and are easily stripped by the cylinder wire clothing. This implies that the transfer involves a reversal of the leading and trailing ends of the fibres. Further orientation and parallelism of the fibre mass is thought to occur during the transfer onto the cylinder.

No experimental work has been published which specifically involves a study of the transfer of fibres from the taker-in to the cylinder. Therefore it has yet to be established whether at the interface, the cylinder, which has the faster surface speed, strips the fibre mass with its clothing or the taker-in, through the action of centrifugal forces, ejects the tuftlets and single fibres onto the cylinder, or a combination of both occurs. It is also of interest to determine if the airflow in the region assists the fibre mass transfer. Whatever the case, the fibre mass is likely to be subjected to an uncontrolled drafting effect, which could introduce irregularities in the mass flow.

**Zone 2: The Fibre Carding Zone**

In the carding zone, it is the interaction of the fibre mass and the wire-teeth clothing of cylinder and flats that fully individualises the fibres and gives parallelism to the fibre mass flow.

In considering how fibres enter and are individualised in the carding zone, Oxley suggests that tuftlets are not strongly held on the cylinder clothing because the tooth angle faces the direction of cylinder rotation. They are, thus, easily removed and more firmly held by the opposing teeth of the flats. It is therefore assumed that as a flat enters the carding zone it becomes almost fully loaded with fibres, the airflow within the region assisting the fibre mass transfer. Having been stripped of fibre mass, subsequent following areas of the cylinder wire clothing move past the fully loaded flat and proceed to comb fibres from the flat, carrying them towards the doffer. The action of combing causes the fibres to be hooked around the cylinder wire points and prevents them from being easily removed by other flats. Debar and Watson's experiments of the movement of radioactive tracer fibres through a miniature card showed that some fibres caught by the flats were often only removed by the cylinder-wire clothing after many revolutions of the cylinder.

Varga reports an alternative view to Oxley's, stating that two types of action occur at the cylinder-flats interface. First, a carding action where the upper layer of a tuftlet or a loosely opened fibre group is caught and held by the flats whilst simultaneously the bottom layer is sheared away by the fast moving cylinder surface. This action causes the top to hang from the flats and to contact subsequent parts of the cylinder wire surface resulting in the second action which is combing, where the wire clothing of the cylinder hooks single or a small group of fibres and combs them from the top layer. A second flat catches the bottom layer on the cylinder and the actions are repeated. In this way tuftlets or groups of fibres are separated into individual fibres.

By making abrupt changes in the colour of the fibre mass fed to the card, Oxley demonstrated that tuftlets from the load on a given flat are carried forward by the cylinder clothing and separated into individual fibres over a small number of preceding flats, typically 4. It was concluded that the interchange of fibres between cylinder and flats does not occur over the full carding zone. Sengupta and made measurements of the carding/combing forces and showed that essentially these actions were on average confined to the first ten working flats. A study by Hodgson showed that moving in the direction of the cylinder rotation, a given flat acquires two-thirds of its final load directly it comes into position over the cylinder.

The load then increases exponentially with time, reaching nine-tenths of the final value within 6-8 minutes. Completion of the load takes place slowly during the remainder of the working time.

The flats moving in the reverse direction the load first increases rapidly with time and then slows until the flat is about to leave the working area. Here it encounters the fibre layer being transported on the cylinder surface from the taker-in. The flat receives a sudden addition of fibre mass to become fully loaded, and, in agreement with other results], the load weighs more than for the forward direction of motion. Contrary to Oxley's conclusions, it was found that 30% of the final load on a given flat resulted from fibre interchange between flats and cylinder over the full carding zone.

It may be reasoned that the number of flats involved in separating a tuftlet depends on the tuftlet size, the mass flow rate and the flat setting. Large tuftlets will be pressed into the cylinder wire during the carding action, whereas small tuftlets will be more easily carded and will remain at the top of the cylinder wire teeth. The larger the tuftet, the higher the production rate and the closer the flat settings, the greater the number flats involved in the separation of a given tuftlet.

Bogdan reports that flats tend to load quickly at the beginning of their cycle of contact with the cylinder. This, however, is only a partial loading, since the fibre mass tends to resist more fibres entering the space but, in the case of cotton, not the leaf and trash particles present.

Analysis of the trash in cotton flat strips showed that initially the percentage of trash in a given flat strip is low and increases slowly during the first 10 minutes of carding, then remains at almost a constant value.

The final percentage depends on the trash content of the cotton. For a fixed production rate, the amount of flat strips was found to be directly proportional to the flat speed, but provided the speed was such that the working time was not less than 10 minutes, both the weight and composition of the flat strips remained approximately constant.

Feil claims that a high degree of air turbulence exists in the flat/cylinder zone. A combination of centrifugal forces, mechanical contact with the flat wire and air turbulence causes the trailing ends of fibres attached individually to the cylinder clothing to vibrate and shake loose trash and dust particles. Short fibres which cannot adequately cling to the cylinder clothing will also be shaken free, and along with impurities become part of the flat strips.

Fibres that are deeply embedded in the flats, and cannot be reached by the cylinder wires become flat strips. For this reason the closeness of the flats setting to the cylinder is important. It may be assumed that closer flats/cylinder setting and faster cylinder speeds will give more effective carding and combing actions as described by Varga and thereby improve web quality through reduced neps and trash.

Cylinder diameters vary and Karasev showed mathematically that for a given cylinder rotational speed the carding power will be greater for a larger cylinder diameter with a higher number of working flats. However, because of lower mechanical stresses, smaller cylinders can be rotated at higher speeds than larger cylinders. The above advantage is therefore reduced the higher the speed of the smaller cylinder.

Artzt and et al studying the influence of card clothing parameters and cylinder speeds on yarn imperfections, report that the teeth density of the flats and cylinder, and the speed of the cylinder must prevent tuftlets lying within the spiral pitch of the cylinder clothing. If this occurs the tuftlets generally become the thick places in the yarn. It was found that high teeth densities and low cylinder speeds were as effective as lower teeth densities and high cylinder speed. High teeth densities with high cylinder speeds did not give effective carding, but no reason was reported for this.

Since the action of the cylinder in this region is to individualise fibres, the wire clothing on the cylinder has a steeper rake and a higher point density than the wire clothing of the flats.

Thus, with closer settings and higher cylinder speeds greater forces may be involved and may result in fibre breakage. However, the work of Li and et al indicates that the withdrawal forces needed to separate an entangled fibre mass was largely dependent on the density of the fibre mass and the contact angle fibres made with the wire clothing, than on the machine speeds.

Van Alphen reports that increasing cylinder speed causes more fibre breakage than increasing taker-in speed and that this is reflected in the yarn properties.

Rotor yarn tenacity was reduced by up to 5% with increasing cylinder speeds between 480 -600 r/min. Whereas ring yarns showed a 5% reduction for speeds between 260 - 380 r/min and 10% at 600 r/min. The higher sensitivity of ring yarns to fibre breakage was attributed to the negative effect of short fibres during roller drafting. Krylov reports that no fibre shortening was observed for cylinder speeds up to 380 r/min.

It may be reasoned that the smaller the tuftlets and the more parallel fibres in tuftlets are to the direction of mass flow the lower the probability of fibre breakage. Honold attributes fibre damage to the cylinder/flat interaction and suggested that the degree of damage depends on the size of the tuftlets entering the working area; the smaller the tuftlets, the closer the setting that can be used and the lower the fibre breakage.

Hodgson's work showed fibre length is also an important factor. For cottons, fibre breakage was only found to have occurred when the staple length was greater than 25mm. Increasing the flat speed appears to have no effect on fibre breakage. However, the amount of flat strips increased proportionally with the flat speed and the mean fibre length of the strips increased significantly. This means that faster flat speeds result in larger amounts of useable fibre in the waste. Interestingly, when carding cottons, immature fibres were not readily found in the flat strips. The coarser rigid fibres seem more easily retained by the flats.

The effectiveness of the carding and combing actions within the cylinder/ flats area is, inter alia, dependent on the quantity of fibre mass on the cylinder, and this includes the recycling layer, $Q_2$. It is of interest therefore to consider how the $Q_2$ is formed during fibre transfer from cylinder to doffer, and its importance to the card web quality.

**Zone 3: Cylinder/ Doffer Interaction**

Varga reports that the action of fibre mass transfer to the doffer is similar to the transfer at the input to the cylinder-flats zone. The regions above and below the line of closest approach of the cylinder to the doffer (*i.e.* the setting line) are important to the mechanism of fibre mass transfer and the transfer coefficient, K.

The two regions may be termed the top and bottom co-operation arcs or top and bottom zones. Simpson claims that transfer can occur in both zones and that the particular region in which transfer actually occurs influences the fibre configuration and the nep level of the card web, although cylinder-flats action is more important in reducing neps. Which zone transfer occurs in is dependent on the cylinder-doffer surface speed ratio, C/D.

For high C/Ds, transfer occurs in the top zone and results in a larger number of trailing than leading hook fibres and a low nep level. The reverse occurs when transfer takes place in the bottom zone owing to lower C/Ds. Simpson does not however say at what C/D value transfer changes from one zone to the other. Although reference is made to other authors who have proposed a mechanism for fibre transfer in the top zone, no mechanism or experimental evidence is given to support the idea of fibre transfer in the bottom zone. Lauber and Wulfhorst used laser-doppler anemometry and high-speed cine photography to study fibre behaviour in the bottom zone, *i.e.* up to110 mm below the setting line. Their findings showed no evidence of fibre transfer within the bottom zone.

Since Morton and Summers' work in 1949 other researchers have confirmed that the values given in Table below for the five classes of fibre configuration observed in slivers. It is of interest to note that the hooked lengths are greater for leading than trailing hooks. Although, the calendar draft can be used to change the relative proportions, Gosh and Bhaduri showed that the method of removing the web from the doffer does not influence the propensity of any class of configuration. It is the mechanism of transfer that is seen as principally responsible for the shape fibres have in the sliver.

**Table. Classification of Fibre Configuration in Card Sliver**

| Calender Draft | Group I Leading Hooks % | M.F.E (mm) | Group II Trailing Hooks % | M.F.E (mm) | Group III Both ends Hooked % | M.F.E (mm) | Group IV No Hooks % | M.F.E (mm) | Group V Other % | M.F.E (mm) |
|---|---|---|---|---|---|---|---|---|---|---|
| | 19 | 17.3 | 51.0 | 17.1 | 13 | 18.6 | 16 | 18.9 | 1 | 10.0 |
| 1.07 | 18 | 18.7 | 50.0 | 17.4 | 12 | 16.0 | 16 | 21.4 | 4 | 7.0 |
| 1.11 | 12 | 20.0 | 52.0 | 15.8 | 15 | 13.0 | 19 | 22.3 | 2 | 8.5 |
| 1.25 | 16 | 20.0 | 38.0 | 18.9 | 13 | 14.3 | 32 | 23.5 | 1 | 12.0 |
| Mean | 16.25 | 19.0 | 47.75 | 17.3 | 13.25 | 15.7 | 20.75 | 21.5 | 2 | 9.37 |

Several studies have been reported on the fibre-mass-transfer mechanism. A number used tracer fibres with one end of a fibre dyed a different colour from the other.The reported findings suggest that fibre mass transfer occurs by fibres acting independently and not as a web of fibres. Observations showed that prior to transfer, nearly 70% of fibres on the cylinder had leading hooks, only 9% had trailing hooks. On transfer the relative proportions changed as indicated in Table. Half the number observed underwent reversals, with greater than 70% changing their configurations [*e.g.* leading hooks becoming trailing hooks]. Of those that transferred without reversals ca 90% did so with a change of configuration.

**Table. Mode of Fibre Transfer from Cylinder to Doffer**

| Configuration | On Cylinder (%) | Mode of fibre transfer to doffer: Reversal (%) No change of config. | Change of config. | No reversal (%) No change of config. | Change of config |
|---|---|---|---|---|---|
| Trailing | 4.5 | 2.25 | 2.25 | 0.00 | 0.00 |
| Leading | 63.7 | 15.90 | 13.70 | 4.50 | 29.60 |
| No hooks | 27.3 | 2.20 | 11.45 | 2.20 | 11.45 |
| Both ends hook | 4.5 | 0.00 | 2.25 | 0.00 | 2.25 |
| Total | 100.0 | 20.35 | 29.65 | 6.7 | 43.30 |

Ghosh and Bhaduri report that tracer fibres were noted generally to go around with the cylinder for several revolutions before being transferred by the doffer. On occasions transfer only happened when the cylinder speed was increased. Debar and Watson's work with radioactive viscose tracer fibres showed that a fibre on the cylinder wire passes the doffer up to a maximum 20 times before being removed by the doffer, sometimes interchanging several times between the cylinder and flats, during the 20 revolutions on the cylinder. Hodgson found that cotton fibres make between 10 and 25 cylinder revolutions before being removed by the doffer. With the continuity of fibre mass flow through the card, this means that the doffer web is built up over many cylinder revolutions and that the recycling layer, $Q_2$, is comprised of multiple fractional layers of the fibre mass transferred from taker-in to cylinder during these cylinder revolutions.

Baturin developed equations that showed the importance of tooth angle and teeth density of the cylinder and doffer wires to the value of K and thereby $Q_2$. Other investigators have reported experimental data that verify Baturin's equations. It was found that the more acute the working angle of the doffer wire compared to the cylinder wire, the higher the value of K, and the lower $Q_2$, and that higher teeth densities on the doffer increased K. These findings would tend to suggest that the proposed mechanism is a principal action by which fibres are removed from the cylinder. However, this mechanism of fibre transfer does not explain the change of fibre configuration with reversals and the formation of leading hooks in the doffer web. It also does not explain how fibres forming the recycling layer, $Q_2$, are subsequently removed, even though an input layer of fibre mass is added to $Q_2$ each time it passes the taker-in.

The above studies did not take account of the degree of fibre parallelism on the cylinder prior to transfer, nor the number of fibres per tooth on the cylinder and consequently the likelihood of fibre interaction during transfer. Fujino and Itani used a microscopic technique to observe the orientation of fibres on the cylinder surface above the taker-in and just before the doffer, and in the doffer web. They found that fibres showed the highest degree of parallelism when on the cylinder surface just above the doffer. The degree of parallelism decreases on transfer to the doffer, and further deteriorates when the web is removed from the doffer to form the sliver, even though the calendar draft helps to maintain some degree of parallelism.

Grimshaw and others report the use of fixed flats just before the cylinder/doffer top transfer zone, to improve fibre parallelism in the card web.; up to 20% reduction in fibre hooks and 25% improvement in fibre parallelism were obtained in the card web, resulting in improved yarn properties. The fixed flats in this region are more effective in improving yarn properties compared with the fixed flats above the taker-in. The action of the flats fitted above the doffer is not fully understood. It is assumed that they tend to lift the fibres to the tip of the cylinder wire for more effective transfer to the doffer, particularly at high cylinder speed. Lauber and Wolfhorst, Kamogawa, report that in this region aerodynamic forces affect the parallelism of the fibres and the way they are transferred to the doffer. However, no details are given.

Owing to the higher speed and larger diameter of the cylinder, it is assumed that during transfer in the top zone the fibres are more substantially affected by the flow of air transported with the cylinder's than by the doffer's wire clothing. High-speed photographs showed that in the bottom zone the main flow of fibre mass was with the doffer at close to the doffer speed, even when the fibres were just below the cylinderdoffer setting line. However, some fibres were seen to be free of both the doffer and cylinder and tended to move with the air currents and eventually with the motion of the cylinder surface.

From the above discussion, it can be seen that work is still needed to establish a more detailed understanding of fibre mass transfer between the cylinder and doffer. The results of such work may also help in better explaining how fibres remain on the cylinder to form the recycling layer $Q_{2.}$ Varga suggests that with fibre transfer in the top zone, the thicker layer of web on the doffer surface protrudes above the doffer wire and into the gap setting between doffer and cylinder. The faster moving cylinder wire clothing combs through the doffer web and thereby pulls fibres back onto the cylinder surface. De Swann showed that fibres can be readily transferred from the doffer to the cylinder as well as from cylinder to doffer. In Hodgson's study, changing cylinder/doffer setting affected the neppiness of the web but did not affect K, which seems to contradict Varga's view. Baturin and Simpson however showed that K will increase if the region of interaction between the cylinder and doffer is reduce by decreasing the doffer or the cylinder diameter and this tend to supports Varga's suggestion for a combing and robbing action of the cylinder. It is reasonable to assume that the combing action could lead to fibres in Class II and IV, but there is still no verified explanation of how fibres in Classes I, III, and V are formed, with and without reversals.

Much of the research on the cylinder/ doffer interaction concerns the effect of machine variables on the size of $Q_2$ (or the operational layer, $Q_0$), on the web quality and changes to the relative proportions of the classified configurations, and on ultimately the yarn quality.

Sing and Swani developed a Markovian model for the carding process in order to determine the probabilities of fibre transfer between cylinder and flats and cylinder and doffer, taking into account the recycling of fibres. It was shown that the times spent by a fibre on the cylinder, $T_{r,}$ and in the flats/cylinder region, $T_{d,}$ are given by:

$$Tr = 1/ K \text{ and } T_d = T_{r.} P_f$$

Where $K = Q_{1/} Q_0$ and $P_f = Q_{f/} Q_0$

Reported values for K would seem to vary between 0.2% to 20%, depending on doffer and cylinder speeds, on the relative profiles of the saw-tooth wire clothing, and on the sliver count. Simpson suggests that fibre properties are also of importance, in that there is a tendency for low micronaire cottons to give higher cylinder loading and for fibres with low shear friction and good compression recovery to result in higher K values. No physical explanation is given for these findings and no other studies are reported on the effect of fibre properties. Further work is therefore needed in this area.

A popular view is that a low fibre mass entering the cylinder/flats interface, *i.e.* a low fibre load on the cylinder, results in better quality carding. This would seem to imply that the higher the value of K the better the carding since less fibre mass is recycling to be added to the mass transferred from the taker-in. However, there are several ways of increasing K and not all of them result in improved carding quality. The cylinder speed and sliver count, increased doffer speed increases K and reduces $P_{f,}$ whereas keeping the doffer speed and sliver count constant and increasing the cylinder speed increase both K and $P_f$. For constant cylinder and doffer speeds, increased sliver count was found to reduce K and $P_f$.

If the same up stream machinery is used, then the best measure of effective carding is the quality of the carded ring-spun yarns produced. Gosh and Bhaduri's work showed that for a fixed carding rate, with increasing doffer or cylinder speed, K increases but $Q_o$ and the yarn imperfections decrease; no trend was found with yarn tenacity or irregularity. Singh and Swani studied the properties of yarns made from slivers corresponding to differing K and $P_f$ values and found that $P_f$ was the more important of the two parameters, in that the higher the value of $P_f$ the better the yarn quality. Kaufman reports that the lighter the fibre load is on the flats, the better the carding quality. Thus, the use of $P_f$ does not give an adequate understanding of the importance of the recycling layer nor of the size of the fibre mass load at the cylinder/flats interface.

Baturin reports an alternative approach to the above in which the following expression was derived for the number of cycles, $N_p$, under steady state conditions that fibres on the cylinder clothing make pass the flats before being removed by the doffer:

$$Np = 1 + Vc/KVd$$

Where K is the transfer coefficient

Vc and Vd are cylinder and doffer surface speeds (m/min).

Since this gives the number of times the recycling fibre mass is subjected to the carding action, it may be a better indication than $P_f$ of the importance of $Q_2$. From the expression, $N_p$ decreases when K increases by increasing doffer speed. The constant production rate, web quality decreases when $N_p$ decreases with doffer speed, even though the cylinder load decreases and a high number of cylinder teeth per fibre is obtained. The last two parameters are usually taken as indicative of good carding. The effect of increased doffer speed and sliver count on web quality and there is a consistent trend which suggests that increasing the production rate by increasing the sliver count, instead of doffer speed, gives better web quality. With regard to sliver irregularity, several investigators report theoretical and experimental studies showing that increasing the recycling layer, $Q_2$, reduces the short-term irregularity.

Karasev attempted to show experimentally the importance of $Q_2$ by removing it during carding using a suction extractor. It was found that without $Q_2$ a large proportion of the fibre mass transferred from the takerin became embedded into the empty teeth of the cylinder clothing. Only the larger tuftlets and groups of individual fibres would then be subjected to the carding and combing actions.

Hence, there is a greater chance of small groups of entangled fibres being removed by the doffer. $Q_2$ therefore acts as a support to new layers of fibre mass being transferred form the taker-in, keeping the new fibre mass at the tips of the cylinder wire teeth and thereby promoting the interaction of tuftlets with the flats and cylinder clothing. This idea, however, does not facilitate an explanation of the mechanism by which fibres leave the recycling layer to form part of the doffer web, $Q_1$.

Gupta and suggest that the rotating cylinder could be considered as a large centrifuge that would cause fibres, impurities and seed fragments to migrate to the cylinder periphery and thereby make contact with the flats clothing and, presumably, the doffer teeth. However, no experimental verification of this hypothesis is reported.

Many of the authors have reported the effect of machine variables on fibre configurations within the card sliver and several have related yarn properties to the observed configurations. Generally it was found that for a fixed sliver count increasing the carding rate by increasing the doffer speed, increased the number of minority hooks and reduced the number of majority hooks, irrespective of cylinder speed.

However, for a given doffer speed, increased cylinder speed gave the reverse trend for minority hooks, but no clear trend for majority hooks. Baturin and Brown showed that increased cylinder speed decreases cylinder load owing to the effect of centrifugal forces and Simpson showed that increased cylinder speed also increased minority hooks and decreased majority hooks. Bhaudri reports that when the fibres are forced nearer the surface of the cylinder teeth, either by increasing the fibre load or increasing the centrifugal force on the cylinder, the proportion of minority hooks increases. Simpson found that there was a direct relation between yarn imperfections and increased occurrence of minority hooks and that spinning end breakage rates and yarn imperfection increased with increased card production speed owing to minority hooks. Gosh and Simpson found that heavier slivers had fewer minority hooks. However, the increased draft needed to process the heavier slivers into yarn led to increased yarn imperfections.

## HIGH STRENGTH MATERIALS OF CARBON FIBRES

Carbon fibers are a new breed of high-strength materials. Carbon fiber has been described as a fiber containing at least 90% carbon obtained by the controlled pyrolysis of appropriate fibers. The existence of carbon fiber came into being in 1879 when Edison took out a patent for the manufacture of carbon filaments suitable for use in electric lamps.

However, it was in the early 1960s when successful commercial production was started, as the requirements of the aerospace industry - especially for military aircraft - for better and lightweight materials became of paramount importance. In recent decades, carbon fibers have found wide application in commercial and civilian aircraft, recreational, industrial, and transportation markets.

Carbon fibers are used in composites with a lightweight matrix. Carbon fiber composites are ideally suited to applications where strength, stiffness, lower weight, and outstanding fatigue characteristics are critical requirements. They also can be used in the occasion where high temperature, chemical inertness and high damping are important. The suppliers of *Advanced Composites Materials Association* released 1997 industry statistics on worldwide shipments of carbon fibers for composites. However, from 1997 to 1999 there was a global slowing of carbon fiber demand. According to Mitsubishi Rayon Co. Ltd. (Tokyo, Japan), a carbon fiber producer, worldwide consumption for sporting goods is nearly 11 million lb of carbon fiber

**Table. Worldwide shipment of carbon fibers for composites**

| YEAR | POUNDS |
|---|---|
| 1992 | 13,000,812 |
| 1993 | 14,598,544 |
| 1994 | 17,425,452 |
| 1995 | 19,714,671 |
| 1996 | 20,672,741 |
| 1997 | 25,900,000 |

Currently, the United States of America uses nearly 60% of the world production of carbon fibers and the Japanese account for almost 50% of the world capacity for production. The largest producer of this fiber is *Toray Industries* of Japan. The world production capacity of pitch-based carbon fiber is almost totally based in Japan.

**Table. Us composite Shipment in 1998**

| Market | Percent Of Total Volume |
|---|---|
| Transportation | 31.6 |
| Construction | 20.8 |
| Corrosion- resistant | 11.8 |
| Marine | 10.1 |
| Electrical/Electronics | 10.0 |

| | |
|---|---|
| Consumers | 6.3 |
| Appliances/Business equipment | 5.5 |
| Aircraft | 0.6 |
| Others | 3.3 |

**Classification And Types**

Based on modulus, strength, and final heat treatment temperature, carbon fibers can be classified into the following categories:

Ultra-high-modulus, type UHM (modulus >450Gpa)

- High-modulus, type HM (modulus between 350-450Gpa)
- Intermediate-modulus, type IM (modulus between 200-350Gpa)
- Low modulus and high-tensile, type HT (modulus < 100Gpa, tensile strength > 3.0Gpa)
- Super high-tensile, type SHT (tensile strength > 4.5Gpa)

PAN-based carbon fibers:

- Pitch-based carbon fibers
- Mesophase pitch-based carbon fibers
- Isotropic pitch-based carbon fibers
- Rayon-based carbon fibers
- Gas-phase-grown carbon fibers

Type-I, high-heat-treatment carbon fibers (HTT), where final heat treatment temperature should be above 2000°C and can be associated with high-modulus type fiber.

- Type-II, intermediate-heat-treatment carbon fibers (IHT), where final heat treatment temperature should be around or above 1500°C and can be associated with high-strength type fiber.
- Type-III, low-heat-treatment carbon fibers, where final heat treatment temperatures not greater than 1000°C. These are low modulus and low strength materials.

**MANUFACTURE**

In *Textile Terms and Definitions*, carbon fiber has been described as a fiber containing at least 90% carbon obtained by the controlled pyrolysis of appropriate fibers. The term "graphite fiber" is used to describe fibers that have carbon in excess of 99%. Large varieties of fibers called precursors are used to produce carbon fibers of different morphologies and different specific characteristics.

The most prevalent precursors are polyacrylonitrile (PAN), cellulosic fibers (viscose rayon, cotton), petroleum or coal tar pitch and certain phenolic fibers. Carbon fibers are manufactured by the controlled pyrolysis of organic precursors in fibrous form. It is a heat treatment of the precursor that removes the oxygen, nitrogen and hydrogen to form carbon fibers. It is well established in carbon fiber literature that the mechanical properties of the carbon fibers are improved by increasing the crystallinity and orientation, and by reducing defects in the fiber. The best way to achieve this is to start with a highly oriented precursor and then maintain the initial high orientation during the process of stabilization and carbonization through tension.

## CARBON FIBERS FROM POLYACRYLONITRILE (PAN)

There are three successive stages in the conversion of PAN precursor into high-performance carbon fibers. Oxidative stabilization: The polyacrylonitrile precursor is first stretched and simultaneously oxidized in a temperature range of 200-300°C. This treatment converts thermoplastic PAN to a non-plastic cyclic or ladder compound. $Carbonization: After oxidation, the fibers are carbonized at about 1000°C without tension in an inert atmosphere (normally nitrogen) for a few hours. During this process the non-carbon elements are removed as volatiles to give carbon fibers with a yield of about 50% of the mass of the original PAN. Graphitization: Depending on the type of fiber required, the fibers are treated at temperatures between 1500-3000°C, which improves the ordering, and orientation of the crystallites in the direction of the fiber axis.

## CARBON FIBERS FROM RAYON

Stabilization: Stabilization is an oxidative process that occurs through steps. In the first step, between 25-150°C, there is physical desorption of water. The next step is a dehydration of the cellulosic unit between 150-240°C. Finally, thermal cleavage of the cyclosidic linkage and scission of ether bonds and some C-C bonds via free radical reaction (240-400° C) and, thereafter, aromatization takes place. Carbonization: Between 400 and 700°C, the carbonaceous residue is converted into a graphite-like layer. Graphitization: Graphitization is carried out under strain at 700-2700°C to obtain high modulus fiber through longitudinal orientation of the planes.

The carbon fiber fabrication from pitch generally consists of the following four steps:

Pitch preparation: It is an adjustment in the molecular weight, viscosity, and crystal orientation for spinning and further heating.

Spinning and drawing: In this stage, pitch is converted into filaments, with some alignment in the crystallites to achieve the directional characteristics.

Stabilization: In this step, some kind of thermosetting to maintain the filament shape during pyrolysis. The stabilization temperature is between 250 and 400 °C. Carbonization: The carbonization temperature is between 1000-1500°C.

## CARBON FIBERS IN MELTBLOWN NONWOVENS

Carbon fibers made from the spinning of molten pitches are of interest because of the high carbon yield from the precursors and the relatively low cost of the starting materials. Stabilization in air and carbonization in nitrogen can follow the formation of melt-blown pitch webs. Processes have been developed with isotropic pitches and with anisotropic mesophase pitches. The mesophase pitch based and melt blown discontinuous carbon fibers have a peculiar structure. These fibers are characterized in that a large number of small domains, each domain having an average equivalent diameter from 0.03mm to 1mm and a nearly unidirectional orientation of folded carbon layers, assemble to form a mosaic structure on the cross-section of the carbon fibers. The folded carbon layers of each domain are oriented at an angle to the direction of the folded carbon layers of the neighboring domains on the boundary.

## CARBON FIBERS FROM ISOTROPIC PITCH

The isotropic pitch or pitch-like material, *i.e.*, molten polyvinyl chloride, is melt spun at high strain rates to align the molecules parallel to the fiber axis. The thermoplastic fiber is then rapidly cooled and carefully oxidized at a low temperature (<100°C). The oxidation process is rather slow, to ensure stabilization of the fiber by cross-linking and rendering it infusible. However, upon carbonization, relaxation of the molecules takes place, producing fibers with no significant preferred orientation. This process is not industrially attractive due to the lengthy oxidation step, and only low-quality carbon fibers with no graphitization are produced. These are used as fillers with various plastics as thermal insulation materials.

## CARBON FIBERS FROM ANISOTROPIC MESOPHASE PITCH

High molecular weight aromatic pitches, mainly anisotropic in nature, are referred to as mesophase pitches. The pitch precursor is thermally treated above 350°C to convert it to mesophase pitch, which contains both isotropic and anisotropic phases. Due to the shear stress occurring during spinning, the mesophase molecules orient parallel to the fiber axis. After spinning, the isotropic part of the pitch is made infusible by thermosetting in air at a temperature below it's softening point. The fiber is then carbonized at temperatures up to 1000°C. The main advantage of this process is that no tension is required during the stabilization or the graphitization, unlike the case of rayon or PAN precursors.

The characterization of carbon fiber microstructure has been mainly been performed by x-ray scattering and electron microscopy techniques. In contrast to graphite, the structure of carbon fiber lacks any three dimensional order. In PAN-based fibers, the linear chain structure is transformed to a planar structure during oxidative stabilization and subsequent carbonization. Basal planes oriented along the fiber axis are formed during the carbonization stage. Wide-angle x-ray data suggests an increase in stack height and orientation of basal planes with an increase in heat treatment temperature. A difference in structure between the sheath and the core was noticed in a fully stabilized fiber. The skin has a high axial preferred orientation and thick crystallite stacking. However, the core shows a lower preferred orientation and a lower crystallite height.

In general, it is seen that the higher the tensile strength of the precursor the higher is the tenacity of the carbon fiber. Tensile strength and modulus are significantly improved by carbonization under strain when moderate stabilization is used. X-ray and electron diffraction studies have shown that in high modulus type fibers, the crystallites are arranged around the longitudinal axis of the fiber with layer planes highly oriented parallel to the axis. Overall, the strength of a carbon fiber depends on the type of precursor, the processing conditions, heat treatment temperature and the presence of flaws and defects. With PAN based carbon fibers, the strength increases up to a maximum of 1300°C and then gradually decreases. The modulus has been shown to increase with increasing temperature. PAN based fibers typically buckle on compression and form kink bands at the innermost surface of the fiber. However, similar high modulus type pitch-based fibers deform by a shear mechanism with kink bands formed at 45° to the fiber axis. Carbon fibers are very brittle. The layers in the fibers are formed by strong covalent bonds. The sheet-like aggregations allow easy crack propagation. On bending, the fiber fails at very low strain.

The two main applications of carbon fibers are in specialized technology, which includes aerospace and nuclear engineering, and in general engineering and transportation, which includes engineering components such as bearings, gears, cams, fan blades and automobile bodies. Recently, some new applications of carbon fibers have been found. Such as rehabilitation of a bridge in building and construction industry.

Others include: decoration in automotive, marine, general aviation interiors, general entertainment and musical instruments and after-market transportation products. Conductivity in electronics technology provides additional new application. Some of the characteristics and applications of carbon fibers.

The production of highly effective fibrous carbon adsorbents with low diameter, excluding or minimizing external and intra-diffusion resistance to mass transfer, and therefore, exhibiting high sorption rates is a challenging task. These carbon adsorbents can be converted into a wide variety of textile forms and nonwoven materials. Cheaper and newer versions of carbon fibers are being produced from new raw materials. Newer applications are also being developed for protective clothing (used in various chemical industries for work in extremely hostile environments), electromagnetic shielding and various other novel applications. The use of carbon fibers in Nonwovens is in a new possible application for high temperature fire-retardant insulation (e.g.: furnace material.)

# CARBON-FIBER-REINFORCED POLYMER

Carbon-fiber-reinforced polymer or carbon-fiber-reinforced plastic (CFRP or CRP or often simply carbon fiber), is an extremely strong and lightfiber-reinforced polymer which contains carbon fibers. The polymer is most often epoxy, but other polymers, such as polyester, vinyl ester or nylon, are sometimes used. The composite may contain other fibers, such as Kevlar, aluminium, or glass fibers, as well as carbon fiber. The strongest and most expensive of these additives, carbon nanotubes, are contained in some primarily polymer baseball bats, car parts and even golf clubs where economically viable.

Although carbon fiber can be relatively expensive, it has many applications in aerospace and automotive fields, such as Formula One. The compound is also used in sailboats, rowing shells, modern bicycles, and motorcycles, where its high strength-to-weight ratio and very good rigidity is of importance. Improved manufacturing techniques are reducing the costs and time to manufacture, making it increasingly common in small consumer goods as well, such as certain ThinkPads since the 600 series, tripods, fishing rods, hockey sticks, paintball equipment, archery equipment, tent poles, racquet frames, stringed instrument bodies, drum shells, golf clubs, helmets used as a paragliding accessory and pool/billiards/snooker cues.

The material is also referred to as *graphite-reinforced polymer* or *graphite fiber-reinforced polymer* (*GFRP* is less common, as it clashes with glass-(fiber)-reinforced polymer). In product advertisements, it is sometimes referred to simply as *graphite fiber* for short.

## PROPERTIES

Carbon-fiber-reinforced polymers are composite materials. In this case the composite consists of two parts; a matrix and a reinforcement. In CFRP the reinforcement is carbon fiber, which provides the strength. The matrix is usually a polymer resin, such as epoxy, to bind the reinforcements together. Because CFRP consists of two distinct elements, the material properties depend on these two elements. The reinforcement will give the CFRP its strength and rigidity; measured by Stress (mechanics) and Elastic modulus respectively. Unlike isotropic materials like steel and aluminum, CFRP has directional strength properties. The properties of CFRP depend on the layouts of the carbon fiber and the proportion of the carbon fibers relative to the polymer.

### Manufacture

The process by which most carbon-fiber-reinforced polymer is made varies, depending on the piece being created, the finish (outside gloss) required, and how many of this particular piece are going to be produced. In addition, the choice of matrix can have a profound effect on the properties of the finished composite.

### Moulding

One method of producing graphite-epoxy parts is by layering sheets of carbon fiber cloth into a mold in the shape of the final product. The alignment and weave of the cloth fibers is chosen to optimize the strength and stiffness properties of the resulting material. The mold is then filled with epoxy and is heated or air-cured. The resulting part is very corrosion-resistant, stiff, and strong for its weight. Parts used in less critical areas are manufactured by draping cloth over a mold, with epoxy either preimpregnated into the fibers (also known as *pre-preg*) or "painted" over it. High-performance parts using single molds are often vacuum-bagged and/or autoclave-cured, because even small air bubbles in the material will reduce strength.

### Vacuum Bagging

For simple pieces of which relatively few copies are needed, (1–2 per day) a vacuum bag can be used. A fiberglass, carbon fiber or aluminum mold is polished and waxed, and has a release agent applied before the fabric and resin are applied, and the vacuum is pulled and set aside to allow the piece to cure (harden). There are two ways to apply the resin to the fabric in a vacuum mold. One is called a wet layup, where the two-part resin is mixed and applied before being laid in the mold and placed in the bag. The other is a resin induction system, where the dry fabric and mold are placed inside the bag while the vacuum pulls the resin through a small tube into the bag, then through a tube with holes or something similar to evenly spread the resin throughout the fabric. Wire loom works perfectly for a tube that requires holes inside the bag. Both of these methods of applying resin require hand work to spread the resin evenly for a glossy finish with very small pin-holes. A third method of constructing composite materials is known as a dry layup. Here, the carbon fiber material is already impregnated with resin (prepreg) and is applied to the mold in a similar fashion to adhesive film. The assembly is then placed in a vacuum to cure. The dry layup method has the least amount of resin waste and can achieve lighter constructions than wet layup. Also, because larger amounts of resin are more difficult to bleed out with wet layup methods, prepreg parts generally have fewer pinholes. Pinhole elimination with minimal resin amounts generally require the use of autoclave pressures to purge the residual gases out.

### Compression Moulding

A quicker method uses a compression mold. This is a two-piece (male and female) mold usually made out of fiberglass or aluminum that is bolted together with the fabric and resin between the two. The benefit is that, once it is bolted together, it is relatively clean and can be moved around or stored without a vacuum until after curing. However, the molds require a lot of material to hold together through many uses under that pressure.

## Filament Winding

For difficult or convoluted shapes, a filament winder can be used to make pieces.

## Structure

Many carbon-fiber-reinforced polymer parts are created with a single layer of carbon fabric that is backed with fiberglass. A tool called a chopper gun is used to quickly create these composite parts. Once a thin shell is created out of carbon fiber, the chopper gun cuts rolls of fiberglass into short lengths and sprays resin at the same time, so that the fiberglass and resin are mixed on the spot. The resin is either external mix, wherein the hardener and resin are sprayed separately, or internal mixed, which requires cleaning after every use.

The primary element of CFRP is a fiber. From these fibers, a unidirectional sheet is created. These sheets are layered onto each other in a quasi-isotropic layup, *e.g.* 0, +60, "60 degrees relative to each other. From the elementary fiber, a bidirectional woven sheet can be created, *i.e.* a twill with a 2/2 weave.

## Automotive Uses

Carbon-fiber-reinforced polymer is used extensively in high-end automobile racing. The high cost of carbon fiber is mitigated by the material's unsurpassed strength-to-weight ratio, and low weight is essential for high-performance automobile racing. Racecar manufacturers have also developed methods to give carbon fiber pieces strength in a certain direction, making it strong in a load-bearing direction, but weak in directions where little or no load would be placed on the member. Conversely, manufacturers developed omnidirectional carbon fiber weaves that apply strength in all directions. This type of carbon fiber assembly is most widely used in the "safety cell" monocoque chassis assembly of high-performance racecars.

Many supercars over the past few decades have incorporated CFRP extensively in their manufacture, using it for their monocoque chassis as well as other components.

Cast vinyl has also been used in automotive applications for aesthetics, as well as heat and abrasion resistance. Most top-of-the-line cast vinyl materials such as 3M's DiNoc (interior use) and SI's Si-1000 3D (exterior use) have lifespans of 10+ years when installed correctly.

Until recently, the material has had limited use in mass-produced cars because of the expense involved in terms of materials, equipment, and the relatively limited pool of individuals with expertise in working with it. Recently, several mainstream vehicle manufacturers have started to use CFRP in everyday road cars.

Use of the material has been more readily adopted by low-volume manufacturers who used it primarily for creating body-panels for some of their high-end cars due to its increased strength and decreased weight compared with the glass-reinforced polymer they used for the majority of their products.

Use of carbon fiber in a vehicle can appreciably reduce the weight and hence the size of its frame. This will also facilitate designers' and engineers' creativity and allow more in-cabin space for commuters.

## CIVIL ENGINEERING APPLICATIONS

Carbon-fiber-reinforced polymer (CFRP) has over the past two decades become an increasingly notable material used in structural engineering applications. Studied in an academic context as to its potential benefits in construction, it has also proved itself cost-effective in a number of field applications strengthening concrete, masonry, steel, cast iron, and timber structures. Its use in industry can be either for retrofitting to strengthen an existing structure or as an alternative reinforcing (or prestressing) material instead of steel from the outset of a project.

Retrofitting has become the increasingly dominant use of the material in civil engineering, and applications include increasing the load capacity of old structures (such as bridges) that were designed to tolerate far lower service loads than they are experiencing today, seismic retrofitting, and repair of damaged structures. Retrofitting is popular in many instances as the cost of replacing the deficient structure can greatly exceed its strengthening using CFRP.

Applied to reinforced concrete structures for flexure, CFRP typically has a large impact on strength (doubling or more the strength of the section is not uncommon), but only a moderate increase in stiffness (perhaps a 10% increase). This is because the material used in this application is typically very strong (*e.g.*, 3000 MPa ultimate tensile strength, more than 10 times mild steel) but not particularly stiff (150 to 250 GPa, a little less than steel, is typical). As a consequence, only small cross-sectional areas of the material are used. Small areas of very high strength but moderate stiffness material will significantly increase strength, but not stiffness.

CFRP can also be applied to enhance shear strength of reinforced concrete by wrapping fabrics or fibers around the section to be strengthened. Wrapping around sections (such as bridge or building columns) can also enhance the ductility of the section, greatly increasing the resistance to collapse under earthquake loading. Such 'seismic retrofit' is the major application in earthquake-prone areas, since it is much more economic than alternative methods.

If a column is circular (or nearly so) an increase in axial capacity is also achieved by wrapping. In this application, the confinement of the CFRP wrap enhances the compressive strength of the concrete. However, although large increases are achieved in the ultimate collapse load, the concrete will crack at only slightly enhanced load, meaning that this application is only occasionally used. Specialist ultra-high modulus CFRP (with tensile modulus of 420 GPa or more) is one of the few practical methods of strengthening cast-iron beams. In typical use, it is bonded to the tensile flange of the section, both increasing the stiffness of the section and lowering the neutral axis, thus greatly reducing the maximum tensile stress in the cast iron.

When used as a replacement for steel, CFRP bars could be used to reinforce concrete structures, however the applications are not common.

CFRP could be used as prestressing materials due to their high strength. The advantages of CFRP over steel as a prestressing material, namely its light weight and corrosion resistance, should enable the material to be used for niche applications such as in offshore environments. However, there are practical difficulties in anchorage of carbon fiber strands and applications of this are rare.

In the United States, prestressed concrete cylinder pipes (PCCP) account for a vast majority of water transmission mains. Due to their large diameters, failures of PCCP are usually catastrophic and affect large populations. Approximately 19,000 miles of PCCP have been installed between 1940 and 2006. Corrosion in the form of hydrogen embrittlement has been blamed for the gradual deterioration of the prestressing wires in many PCCP lines. Over the past decade, CFRPs have been utilized to internally line PCCP, resulting in a fully structural strengthening system. Inside a PCCP line, the CFRP liner acts as a barrier that controls the level of strain experienced by the steel cylinder in the host pipe. The composite liner enables the steel cylinder to perform within its elastic range, to ensure the pipeline's long-term performance is maintained. CFRP liner designs are based on strain compatibility between the liner and host pipe. CFRP is a more costly material than its counterparts in the construction industry, glass fiber-reinforced polymer (GFRP) and aramid fiber-reinforced polymer (AFRP), though CFRP is, in general, regarded as having superior properties.

Much research continues to be done on using CFRP both for retrofitting and as an alternative to steel as a reinforcing or prestressing material. Cost remains an issue and long-term durability questions still remain. Some are concerned about the brittle nature of CFRP, in contrast to the ductility of steel. Though design codes have been drawn up by institutions such as the American Concrete Institute, there remains some hesitation among the engineering community about implementing these alternative materials. In part, this is due to a lack of standardization and the proprietary nature of the fiber and resin combinations on the market.

**Sporting Goods**

Carbon-fiber-reinforced polymer has found use in high-end sports equipment such as racing bicycles. For the same strength, a carbon fiber frame weighs less than a bicycle tubing of aluminum or steel. The choice of weave can be carefully selected to maximize stiffness. The variety of shapes it can be built into has further increased stiffness and also allowed aerodynamic considerations into tube profiles. Carbon-fiber-reinforced polymer frames,forks, handlebars, seatposts, and crank arms are becoming more common on medium- and higher-priced bicycles. Carbon-fiber-reinforced polymer forks are used on most new racing bicycles. Other sporting goods applications include rackets, fishing rods, longboards, and rowing shells. The large majority of NHL ice hockey players use carbon-fiber sticks. Modern Pole vaulting poles are made of Carbon-fiber. Amputees athletes like Oscar Pistorius use carbon fiber blades for running.

Shoe manufacturers use carbon fiber as a shank plate in their basketball sneakers to keep the foot stable. It usually runs the length of the sneaker just above the sole and is left exposed in some areas, usually in the arch of the foot.

This material is used when manufacturing squash, tennis and badminton racquets. Carbon-Graphite spars are used on the frames of high-end sport kites. In 2006, a company introduced cricket bats with a thin carbon-fiber layer on the back which were used in competitive matches by high-profile players (*e.g.* Ricky Ponting and Michael Hussey). The carbon fiber was claimed to increase the durability of the bats, however they were banned from all first-class matches by the ICC in 2007.

Carbon fiber is used in the manufacture of high quality arrows for archery.

**Aerospace Engineering**

Much of the fuselage of the new Boeing 787 Dreamliner and Airbus A350 XWB will be composed of CFRP, making the aircraft lighter than a comparable aluminum fuselage, with the added benefit of less maintenance thanks to CFRP's superior fatigue resistance. Due to its high ratio of strength to weight, CFRP is widely used in micro air vehicles (MAVs). In the MAVSTAR Project, the CFRP structures reduce the weight of the MAV significantly. In addition, the high stiffness of the CFRP blades overcome the problem of collision between blades under strong wind.

Especially aircraft of the sub-category of microlights (SSDR) take advantage of CFRP. Those aircraft, such as the E-Go, must comply with a weight less than 115 kg (254 lb) without fuel and pilot. Creating an aircraft of this class without the usage of CFRP would be really difficult.

# 6

# Textile Weaving Technology

## WEAVING

Weaving is a method of interlacing two sets of threads, the warp threads and the weft (horizontal) threads, to make cloth. The word textile is often used in place of cloth. A person who makes woven cloth is called a weaver.

### THE DEVELOPMENT OF STRING

20,000 to 30,000 years ago early man developed the first string by twisting together handfuls of plant fibres. Preparing thin bundles of plant material and stretching them out while twisting them together produced a fine string or thread.

The ability to produce string and thread was the starting place for the development of weaving, spinning, and sewing.

### EARLY STONE AGE CLOTH

Stone Age Man's early experiments with string and thread lead to the first woven textiles. Threads and strings of different sizes were knotted and laced together to make many useful articles.

Finger weaving, the lacing and knotting together of threads by hand, is still used today by many weavers.

### NEOLITHIC ERA

During the early Neolithic Era simple weaving looms were developed. Simple weaving looms are man made tools to hold the warp (vertical) threads snugly in order allowing the weaver to insert the weft threads. The two early weaving looms are the horizontal ground loom and the warp weighted loom. [SB] {WW}

### THE WARP WEIGHTED LOOM

This loom is made from large wooden poles tied together in a rectangular shape. The poles can be mounted on a wall or dug into the ground to make a freestanding loom. The warp (vertical) threads are tied to the top pole. At the bottom of the frame the warp threads are tied together in groups and secured to clay or stone weights.

The weaver places the weft threads through the warp by hand while standing in front of the loom. The warp weighted loom is still used today.

### THE HORIZONTAL GROUND LOOM

The horizontal ground loom is a simple arrangement of sticks and poles driven into the ground. The weaver measures out the length and width needed for weaving the cloth and drives the sticks firmly into the ground.

The warp (vertical) threads are wound onto the sticks and tied in place. The weaver works the weft (horizontal) threads, by hand, through the stretched out warp.

The ground loom is still used today by the Bedouin weavers of the Near East.

### NEOLITHIC WOVEN CLOTH

During the Neolithic Era mankind developed great skill in weaving cloth. Every household produced cloth for their own needs.

Weaving cloth remained an activity associated with the family unit for thousands of years. [SB] {WW}

## WEAVING (MYTHOLOGY)

The theme of weaving in mythology is ancient, and its lost mythic lore probably accompanied the early spread of this art. In traditional societies today, westward of Central Asia and the Iranian plateau, weaving is a mystery within woman's sphere. Where men have become the primary weavers in this part of the world, it is possible that they have usurped the archaic role: among the gods, only goddesses are weavers. Herodotusnoted, however, the cultural difference between gender identities and weaving among Hellenes and Egyptians: among Egyptians it was the men who wove.

Weaving begins with spinning. Until the spinning wheel was invented in the 14th century, all spinning was done with distaff and spindle. In English the "distaff side" indicates relatives through one's mother, and thereby denotes a woman's role in the household economy. In Scandinavia, the stars of Orion's belt are known as *Friggjar rockr*, "Frigg's distaff". The spindle, essential to the weaving art, is recognizable as an emblem of security and settled times in a ruler's eighth-century BCE inscription atKaratepe:

"In those places which were formerly feared, where a man fears.. to go on the road, in my days even women walked with spindles"

In the adjacent region of North Syria, historian Robin Lane Fox remarks funerary stelae showing men holding cups as if feasting and women seated facing them and holding spindles.

**Egypt**

In pre-Dynastic Egypt, *nt* (Neith) was already the goddess of weaving (and a mighty aid in war as well). She protected the Red Crown of Lower Egypt before the two kingdoms were merged, and in Dynastic times she was known as the most ancient one, to whom the other gods went for wisdom. Nit is identifiable by her emblems: most often it is the loom's shuttle, with its two recognizable hooks at each end, upon her head. According to E. A. Wallis Budge (*The Gods of the Egyptians*) the root of the word for *weaving* and also for *being* are the same: *nnt*.

**Greece**

In Greece the Moirai (the "Fates") are the three crones who control destiny, and the matter of it is the art of spinning the thread of life on the distaff. Ariadne, the wife of the god Dionysus in Minoan Crete, possessed the spun thread that led Theseus to the centre of the labyrinth and safely out again.

Among the Olympians, the weaver goddess is Athena, who punished the impious pretensions of her acolyte Arachne by turning her into a weaving spider. The daughters of Minyas, Alcithoe,Leuconoe and their sister, defied Dionysus and honored Athena in their weaving instead of joining his festival. A woven peplum, laid upon the knees of the goddess's iconic image, was central to festivals honoring both Athena at Athens, and Hera.

In Homer's legend of the Odyssey, Penelope the faithful wife of Odysseus was a weaver, weaving her design for a shroud by day, but unravelling it again at night, to keep her suitors from claiming her during the long years while Odysseus was away. Penelope has a high lineage that melds human and divine, and is she perhaps secretly Odysseus' own weaving goddess-nymph, like the two weaving enchantresses in the *Odyssey*, Circe and Calypso. Helen is at her loom in the *Iliad*.

Homer dwells upon the supernatural quality of the weaving in the robes of goddesses, and every writer reaching for a heroic style after him imitated an analogous passage.

In the terrible tale of Philomela, who was raped and her tongue cut out so that she could not tell about her violation, her loom becomes her voice, and the story is told in the design, so that her sister Procne may understand and the women may take their revenge. Ovid retold the old tales in his *Metamorphoses* (VI, 575–587). The understanding in the Philomela myth that pattern and design convey myth and ritual has been of great use to modern mythographers: Jane Ellen Harrison led the way, interpreting the more permanent patterns of vase-painting, since the patterned textiles had not survived.

The concept of weaving actually relates to mythology much more than simply appearing in myths, the English word text is derived from the Latin word for weaving, texare, explaining the source of terms like "weaving a story".

**Germanic**

For the Norse peoples, Frigg is a goddess associated with weaving. The Scandinavian "Song of the Spear", quoted in "Njals Saga", gives a detailed description of Valkyries as women weaving on a loom, with severed heads for weights, arrows for shuttles, and human gut for the warp, singing an exultant song of carnage. Ritually deposited spindles and loom parts were deposited with thePre-Roman Iron Age ritual wagon at Dejbjerg, Jutland, and are to be associated with the wagon-goddess.

In Germanic mythology, Holda (Frau Holle) and Perchta (Frau Perchta, Berchta, Bertha) were both known as goddesses who oversaw spinning and weaving. They had many names.

Holda, whose patronage extends outward to control of the weather, and source of women's fertility, and the protector of unborn children, is the patron of spinners, rewarding the industrious and punishing the idle. Holda taught the secret of making linen from flax. An account of Holda was collected by the Brothers Grimm, as the fairy tale "Frau Holda". Another of the Grimm tales, "Spindle, Shuttle, and Needle", which embeds social conditioning in fairy tale with mythic resonances, rewards the industrious spinner with the fulfillment of her mantra:

*"Spindle, my spindle, haste, haste thee away, and here to my house bring the wooer, I pray."*

*"Spindel, Spindel, geh' du aus, bring den Freier in mein Haus."*

This tale recounts how the magic spindle, flying out of the girl's hand, flew away, unravelling behind it a thread, which the prince followed, as Theseus followed the thread of Ariadne, to find what he was seeking: a bride "who is the poorest, and at the same time the richest". He arrives to find her simple village cottage magnificently caparisoned by the magically-aided products of spindle, shuttle and needle.

Jacob Grimm reported the superstition "if, while riding a horse overland, a man should come upon a woman spinning, then that is a very bad sign; he should turn around and take another way." (*Deutsche Mythologie* 1835, v3.135)

Also the Norns, female giantesses, weavers of fate, belong in this folklore of weaving.

**Celts**

The goddess Brigantia, due to her identification with the Roman Minerva, may have also been considered, along with her other traits, to be a weaving deity.

**French**

Weavers had a repertory of tales: in the 15th century Jean d'Arras, a Northern French tale-teller (*trouvere*), assembled a collection of stories entitled

*Les Évangiles des Quenouilles* ("Spinners' Tales"). Its frame story is that these are narrated among a group of ladies at their spinning.

**Baltic**

In Baltic myth, Saule is the life-affirming sun goddess, whose numinous presence is signed by a wheel or a rosette. She spins the sunbeams. The Baltic connection between the sun and spinning is as old as spindles of the sun-stone, amber, that have been uncovered in burial mounds. Baltic legends as told have absorbed many images from Christianity and Greek myth that are not easy to disentangle.

**Finnish**

The Finnish epic, the *Kalevala*, has many references to spinning and weaving goddesses.

**Later European Folklore**

"When Adam delved and Eve span.." runs the rhyme; though the tradition that Eve span is unattested in Genesis, it was deeply engrained in the medieval Christian vision of Eve. In an illumination from the 13th-century Hunterian Psalter (*illustration. left*) Eve is shown with distaff and spindle.

In later European folklore, weaving retained its connection with magic. Mother Goose, traditional teller of fairy tales, is often associated with spinning. She was known as "Goose-Footed Bertha" or *Reine Pédauque* ("Goose-footed Queen") in French legends as spinning incredible tales that enraptured children.The daughter who, her father claimed, could spin straw into gold and was forced to demonstrate her talent, aided by the dangerous earth-daemon Rumpelstiltskinwas an old tale when the Brothers Grimm collected it. Similarly, the unwilling spinner of the tale *The Three Spinners* is aided by three mysterious old women. In the *Six Swans*, the heroine spins and weaves starwort in order to free her brothers from a shapeshifting curse. *Spindle, Shuttle, and Needle* are enchanted and bring the prince to marry the poor heroine. Sleeping Beauty, in all her forms, pricks her finger on a spindle, and the curse falls on her.

In Alfred Tennyson's poem "The Lady of Shalott", her woven representations of the world have protected and entrapped Elaine of Astolat, whose first encounter with reality outside proves mortal. William Holman Hunt's painting from the poem (*illustration, right*) contrasts the completely pattern-woven interior with the sunlit world reflected in the roundel mirror. On the wall, woven representations of Myth ("Hesperides") and Religion ("Prayer") echo the mirror's open roundel; the tense and conflicted Lady of Shalott stands imprisoned within the brass roundel of her loom, while outside the passing knight sings "'Tirra lirra' by the river" as in Tennyson's poem.

A high-born woman sent as a hostage-wife to a foreign king was repeatedly given the epithet "weaver of peace", linking the woman's art and the familiar role of a woman as a dynastic pawn. A familiar occurrence of the phrase is in the early English poem *Widsith*, who "had in the first instance gone with Ealhild, the beloved weaver of peace, from the east out of Anglen to the home of the king of the glorious Goths, Eormanric, the cruel troth-breaker..."

**Inca**

In Inca mythology, Mama Ocllo first taught women the art of spinning thread.

**China**

- In Tang Dynasty China, the goddess weaver floated down on a shaft of moonlight with her two attendants. She showed the upright court official *Guo Han* in his garden that a goddess's robe is seamless, for it is woven without the use of needle and thread, entirely on the loom. The phrase "a goddess's robe is seamless" passed into an idiom to express perfect workmanship. This idiom is also used to mean a perfect, comprehensive plan.
- The Goddess Weaver, daughter of the Celestial Queen Mother and Jade Emperor, wove the stars and their light, known as "the Silver River" (what Westerners call "The Milky Way Galaxy"), for heaven and earth. She was identified with the star Westerners know as Vega. In a 4,000-year-old legend, she came down from the Celestial Court and fell in love with the mortal Buffalo Boy (or Cowherd), (associated with the star Altair). The Celestial Queen Mother was jealous and separated the lovers, but the Goddess Weaver stopped weaving the Silver River, which threatened heaven and earth with darkness. The lovers were separated, but are able to meet once a year, on the seventh day of the seventh moon.

## CONSTRUCTION OF A PLAIN WEAVE

Plain weave is the simplest interlacing pattern which can be produced. It is formed by alternatively lifting and lowering one warp thread across one weft thread. A plain weave fabric in plan view and warp way and weft way cross-sections through the same fabric. The diagrams are idealized because yarns are seldom perfectly regular and the pressure between the ends and picks tends to distort the shape of the yarn cross-sections unless the fabrics are woven from monofilament yarns or strips of film. The yarns also do not lie straight in the fabric because the warp and weft have to bend round each other when they are interlaced. The wave form assumed by the yarn is called crimp and is referred to in greater.

## Diversity of Plain Weave Fabrics

The characteristics of the cloths woven will depend on the type of fibre used for producing the yarn and whether it is a monofilament yarn, a flat, twisted or textured (multi-)filament yarn or whether it has been spun from natural or manufactured staple fibres. The stiffness of the fabrics and its weavability will also be affected by the stiffness of the raw materials used and by the twist factor of the yarn, that is the number of turns inserted in relation to its linear density. Very highly twisted yarns are sometimes used to produce special features in plain weave yarns. Thc resulting fabrics may have high extensibility or can be semiopaque.

The area density of the fabric can be varied by changing the linear density or count of the yarns used and by altering the thread spacing, which affects the area covered by the yarns in relation to the total area. The relation between the thread spacing and the yarn linear density is called the cover factor. Changing the area density and/or the cover factors may affect the strength, thickness, stiffness, stability, porosity, filtering quality and abrasion resistance of fabrics.

Square sett plain fabrics, that is fabrics with roughly the same number of ends and picks per unit area and warp and weft yarns of the same linear densities are produced in the whole range of cloth area densities and cloth cover factors. Low area density fabrics of open construction include bandages and cheese cloths, light area density high cover factor fabrics include typewriter ribbons and medical filter fabrics, heavy open cloths include geotextile stabilization fabrics and heavy closely woven fabrics include cotton awnings.

Warp-faced plain fabrics generally have a much higher warp cover factor than weft cover factor. If warp and weft yarns of similar linear density are used, a typical warp faced plain may have twice as many warp ends as picks. In such fabrics the warp crimp will be high and the weft crimp extremely low. The plan view and cross-sections of such a fabric. By the use of suitable cover factors and choice of yarns most of the abrasion on such a fabric can be concentrated on the warp yarns and the weft will be protected.

Weft-faced plains are produced by using much higher weft cover factors than warp cover factors and will have higher weft than warp crimp. Because of the difference in weaving tension the crimp difference will be slightly lower than in warp-faced plain fabrics. Weft-faced plains are little used because they are more difficult and expensive to weave.

## Construct a Point Paper Diagram

To illustrate a weave either in plan view and/or in cross-section, takes a lot of time, especially for more complicated weaves. A type of shorthand for depicting weave structures has therefore been evolved and the paper used for producing designs is referred to as squared paper, design paper or point paper. Generally the spaces between two vertical lines represent one warp end and the spaces between two horizontal lines one pick. If a square is filled in it represents an end passing over a pick whilst a blank square represents a pick passing over an end. If ends and picks have to be numbered to make it easier to describe the weave, ends are counted from left to right and picks from the bottom of the point paper design to the top. The point paper design is the design for a plain weave fabric. To get a better impression of how a number of repeats would look, four repeats of a design (two vertically and two horizontally) are sometimes shown. When four repeats are shown the first repeat is drawn in the standard way but for the remaining three repeats crossing diagonal lines may be placed into the squares, which in the firsts repeat, are filled in. This method is shown for a plain weave.

## RIB FABRICS AND MATT WEAVE FABRICS

These are the simplest modifications of plain weave fabrics. They are produced by lifting two or more adjoining warp threads and/or two or more adjoining picks at the same time. It results in larger warp and/or weft covered surface areas than in a plain weave fabric. As there are fewer yarn intersections it is possible to insert more threads into a given space, that is to obtain a higher cover factor, without jamming the weave.

### Matt Fabrics (or hopsack)

Simple matt (or hopsack) fabrics have a similar appearance to plain weave. The simplest of the matt weaves is a 2/2 matt, where two warp ends are lifted over two picks, in other words it is like a plain weave fabric with two ends and two picks weaving in parallel. The number of threads lifting alike can be increased to obtain 3/3 or 4/4 matt structures. Special matt weaves, like a 4/2 matt are produced to obtain special technical effects. Larger matt structures give the appearance of squares but are little used because they tend to become unstable, with long floats and threads in either direction riding over each other. If large matt weaves are wanted to obtain a special effect or appearance they can be stabilized by using fancy matt weaves containing a binding or stitching lift securing a proportion of the floats.

Matt weave fabrics can be woven with higher cover factors and have fewer intersections. In close constructions they may have better abrasion and better filtration properties and greater resistance to water penetration. In more open constructions matt fabrics have a greater tear resistance and bursting strength. Weaving costs may also be reduced if two or more picks can be inserted at the same time.

### Rib Fabrics

In warp rib fabrics there are generally more ends than picks per unit length with a high warp crimp and a low weft crimp and vice versa for weft rib fabrics.

The simplest rib fabrics are the 2/2 warp rib and the 2/2 weft rib shown respectively. In the 2/2 warp rib one warp end passes over two picks whilst in the 2/2 weft rib one pick passes over two adjoining ends. The length of the floats can be extended to create 4/4,6/6,3/1 or any similar combination in either the warp or weft direction. 3/1 and 4/4 warp ribs.

In rib weaves with long floats it is often difficult to prevent adjoining yarns from overlapping. Weft ribs also tend to be expensive to weave because of their relatively high picks per unit length which reduces the production of the weaving machine unless two picks can be inserted at the same time.

**Twill Fabrics**

A twill is a weave that repeats on three or more ends and picks and produces diagonal lines on the face of a fabric. Such lines generally run from selvedge to selvedge. The direction of the diagonal lines on the surface of the cloth are generally described as a fabric is viewed along the warp direction. When the diagonal lines are running upwards to the right they are Z twill' or 'twill right' and when they run in the opposite direction they are 'S twill' or 'twill left'. Their angle and definition can be varied by changing the thread spacing and/or the linear density of the warp and weft yarns. For any construction twills will have longer floats, fewer intersections and a more open construction than a plain weave fabric with the same cloth particulars.

Industrial uses of twill fabrics are mainly restricted to simple twills and only simple twills are discussed here. Broken twills, waved twills, herringbone twills and elongated twills are extensively used for suiting and dress fabrics. For details of such twills see Robinson and Marks or Watson. The smallest repeat of a twill weave consists of 3 ends x 3 picks. There is no theoretical upper limit to the size of twill weaves but the need to produce stable fabrics with floats of reasonable length imposes practical limits.

The twill is produced by commencing the lift sequence of adjacent ends on one pick higher or one pick lower. The lift is the number of picks which an end passes over and under. In a 2/1 twill, an end will pass over two picks and under one, whilst in a 1/2 twill the end will pass over one pick and then under two picks. Either weave can be produced as a Z or S twill. There are, therefore, four combinations of this simplest of all twills. To show how pronounced the twill line is even in a 3 x 3 twill, four repeats (2 x 2) lifted ends shown in solid on the point paper. 2/1 twills are warp faced twills, that is fabrics where most of the warp yarn is on the surface, whilst 1/2 twills have a weft face. Weft-faced twills impose less strain on the weaving machine than warp-faced twills because fewer ends have to be lifted to allow picks to pass under them. For this reason, warp-faced twills are sometimes woven upside down, that is as weft-faced twills. Thc disadvantage of weaving twills upside down is that it is more difficult to inspect the warp yarns during weaving.

**Lenos**

In lenos adjoining warp ends do not remain parallel when they are interlaced with the weft but are crossed over each other. In the simplest leno one standard end and one crossing end are passed across each other during consecutive picks. Two variations of this structure are shown in plan view and cross-section. Whenever the warp threads cross over each other, with the weft passing between them, they lock the weft into position and prevent weft movement. Leno weaves are therefore used in very open structures, such as gauzes, to prevent thread movement and fabric distortion. When the selvedge construction of a fabric does not bind its edge threads into position, leno ends are used to prevent the warp threads at each side of a length of cloth from slipping out of the body of the fabric. They are also used in the body of fabrics when empty dents are left in weaving because the fabrics are to be slit into narrower widths at a later stage of processing.

Leno and gauze fabrics may consist of standard and crossing ends only or pairs or multiples of such threads may be introduced according to pattern to obtain the required design. For larger effects standard and crossing ends may also be in pairs or groups of three. Two or more weft threads may be introduced into one shed and areas of plain fabric may be woven in the weft direction between picks where warp ends are crossed over to give the leno effect. Gauze fabrics used for filtration generally use simple leno weaves.

Standard and crossing ends frequently come from separate warp beams. If both the standard and crossing ends are warped on to one beam the same length of warp is available for both and they will have to do the same amount of bending, that is they will have the same crimp. Such a leno fabric is shown in plan view and cross-section.

If the two series of ends are brought from separate beams the standard ends and the crossing ends can be tensioned differently and their crimp can be adjusted separately. In such a case it is possible for the standard ends to lie straight and the crossing ends to do all the bending. That crossing threads can be moved either from the right to the left or from the left to the right on the same pick and adjoining leno pairs may either cross in the same or opposite directions. The direction of crossing can affect locking, especially with smooth monofilament yarns. When using two beams it is also possible to use different types or counts of yarn for standard and crossing ends for design or technical applications. The actual method of weaving the leno with doups or similar mechanisms is not discussed in this chapter.

When lenos are used for selvedge construction only one to four pairs of threads are generally used at each side and the leno selvedge is produced by a special leno mechanism that is independent of the shedding mechanism of the loom. The leno threads required for the selvedge then come from cones in a small separate creel rather than from the warp beam. The choice of selvedge yarns and tensions is particularly important to prevent tight or curly selvedges. The crimps selected have to take into account the cloth shrinkage in finishing.

**Satins and Sateens**

In Britain a satin is a warp-faced weave in which the binding places are arranged to produce a smooth fabric surface free from twill lines. Satins normally have a much greater number of ends than picks per centimetre. To avoid confusion a satin is frequently described as a 'warp satin'. A sateen, frequently referred to as a 'weft sateen', is a weft-faced weave similar to a satin with binding places arranged to produce a smooth fabric free of twill lines. Sateens are generally woven with a much higher number of picks than ends. Satins tend to be more popular than sateens because it is cheaper to weave a cloth with a lower number of picks than ends. Warp satins may be woven upside down, that is as a sateen but with a satin construction, to reduce the tension on the harness mechanism that has to lift the warp ends.

To avoid twill lines, satins and sateens have to be constructed in a systematic manner. To construct a regular satin or sateen weave (for irregular or special ones see Robinson and Marks or Watson ) without a twill effect a number of rules have to be observed. The distribution of interlacing must be as random as possible and there has only to be one interlacing of each warp and weft thread per repeat, that is per weave number. The intersections must be arranged in an orderly manner, uniformly separated from each other and never adjacent. The weaves are developed from a 1/x twill and the twill intersections are displaced by a fixed number of steps. The steps that must be avoided are: (i) one or one less than the repeat (because this is a twill), (ii) the same number as the repeat or having a common factor with the repeat (because some of the yarns would fail to interlace).

The smallest weave number for either weave is 5 and regular satins or sateens also cannot be constructed with weave numbers of 6,9,11,13,14 or 15. The most popular weave numbers are 5 and 8 and weave numbers above 16 are impracticable because of the length of floats. Weave numbers of 2 or 3 are possible for five-end weaves and 3 or 5 for 8-end weaves.

**TRIAXIAL WEAVES**

Nearly all two-dimensional woven structures have been developed from plain weave fabrics and warp and weft yarns are interlaced at right angles or at nearly right angles. This also applies in principle to leno fabrics and to lappet fabrics where a proportion of the warp yarns, that is the yarns forming the design, are moved across a number of ground warp yarns by the lappet mechanism. The only exceptions are triaxial fabrics, where two sets of warp yarns are generally inserted at 60° to the weft, and tetra-axial fabrics where four sets of yarns are inclined at 45° to each other. So far only weaving machines for triaxial fabrics are in commercial production. Triaxial weaving machines were first built by the Barber-Colman Co. under licence from Dow Weave and have been further developed by Howa Machinery Ltd., Japan.

Triaxial fabrics are defined as cloths where the three sets of threads form a multitude of equilateral triangles. Two sets of warp yarn are interlaced at 60° with each other and with the weft. In the basic triaxial fabric, the warp travels from selvedge to selvedge at an angle of 30° from the vertical. When a warp yarn reaches the selvedge it is turned through an angle of 120° and then travels to the opposite selvedge thus forming a firm selvedge. Weft yarn is inserted at right angles (90°) to the selvedge. The basic triaxial fabric is fairly open with a diamond-shaped centre. The standard weaves can be modified by having biplane, stuffed or basic basket weaves, the latter being. These modified weaves form closer structures with different characteristics. Interlacing angles of 75° to the selvedge can be produced. At present thread spacing in the basic weave fabric is limited to 3.6 or 7.4 threads per centimetre.

The tear resistance and bursting resistance of triaxial fabrics is greatly superior to that of standard fabrics because strain is always taken in two directions. Their shear resistance is also excellent because intersections are locked. They have a wide range of technical applications including sailcloths, tyre fabrics, balloon fabrics, pressure receptacles and laminated structures.

**Selvedge**

The selvedges form the longitudinal edges of a fabric and are generally formed during weaving. The weave used to construct the selvedge may be the same, or may differ from, the weave used in the body of the cloth. Most selvedges are fairly narrow but they can be up to 20-mm wide. Descriptions may be woven into the selvedge, using special selvedge jacquards, or coloured or fancy threads may be incorporated for identification purposes. For some end-uses selvedges have to be discarded but, whenever possible, selvedges should be constructed so that they can be incorporated into the final product so as to reduce the cost of waste.

In cloths woven on weaving machines with shuttles selvedges are formed by the weft turning at the edges after the insertion of each pick. The weft passes continuously across the width of the fabric from side to side. In cloths woven on shuttleless weaving machines the weft is cut at the end of each pick or after every second pick. To prevent the outside ends from fraying, various selvedge motions are used to bind the warp into the body of the cloth or edges may be sealed. The essential requirements are that the selvedges should protect the edge of the fabric during weaving and subsequent processing, that they should not detract from the appearance of the cloth and that they should not interfere with finishing or cause waviness, contraction or creasing. Four types of woven selvedges.

## HAIRPIN SELVEDGE - SHUTTLE WEAVING MACHINE

A typical hairpin selvedge which is formed when the weft package is carried in a reciprocating media, for example a shuttle. With most weaves it gives a good edge and requires no special mechanism. Frequently strong two-ply yarns are incorporated into the selvedge whilst single yarns are used in the body of the cloth because the edge threads are subjected to special strains during beat-up. To ensure a flat edge a different weave may be used in the selvedge from the body of the cloth. For twills, satins and fancy weaves this may also be necessary to ensure that all warp threads are properly bound into the edge. If special selvedge yarns are used it is important to ensure that they are not, by mistake, woven into the body of the fabric because they are likely to show up after finishing or cause a reduction in the tear and/or bursting resistance of industrial fabrics.

When two or more different weft yarns are used in a fabric only one weft yarn is being inserted at a time and the other yarn(s) are inactive until the weft is changed. In shuttle looms the weft being inserted at any one time will form a normal selvedge whilst the weft yarn(s) not in use will float along the selvedge. If long floats are formed in weaving, because one weft is not required for a considerable period, the floats may have to be trimmed off after weaving to prevent them from causing problems during subsequent processing. Whenever a pirn is changed, or a broken pick is repaired, a short length of yarn will also protrude from the selvedge and this has to be trimmed off after weaving.

Industrial fabrics with coarse weft yarn are sometimes woven with loop selvedges to ensure that the thread spacing and cloth thickness remains the same right to the edge of the cloth. To produce a loop selvedge a wire or coarse monofilament yarn is placed 3 or 4 mm outside the edge warp end and the pick reverses round the wire to form a loop. As the wire is considerably stiffer than a yarn it will prevent the weft pulling the end threads together during beat-up. The wire extends to the fell of the cloth and is automatically withdrawn during weaving. Great care is required to ensure that no broken-off ends of wire or monofilament remain in the fabric because they can cause serious damage to equipment and to the cloth during subsequent processing.

### Tucked-in Selvedges

A tuck or tucked-in selvedge. It gives a very neat and strong selvedge and was first developed by Sulzer Brothers (now Sulzer-Textil), Win-terthur, Switzerland, for use on their projectile weaving machines. Its appearance is close to that of a hairpin selvedge and it is particularly useful when cloths with fringe selvedges would have to be hemmed. High-speed tucking motions are now available and it is possible to produce tucked-in selvedges even on the fastest weaving machines.

Tucked-in selvedges, even when produced with a reduced number of warp threads, are generally slightly thicker than the body of the cloth. When large batches of cloth with such selvedges are produced it may be necessary to traverse the cloth on the cloth roller to build a level roll. The extra thickness will also have to be allowed for when fabrics are coated. When tucked-in selvedges can be incorporated into the finished product no yarn waste is made because no dummy selvedges are produced but the reduced cost of waste may be counterbalanced by the relatively high cost of the tucking units.

**Sealed Selvedges**

When fabrics are produced from yarns with thermoplastic properties the edge of a fabric may be cut and sealed by heat. The edge ends of fringe selvedges are frequently cut off in the loom and the edge threads with the fringe are drawn away into a waste container. Heat cutting may also be used to slit a cloth, in or off the loom, into a number of narrower fabrics or tapes.

For special purposes ultrasonic sealing devices are available. These devices are fairly expensive and can cut more cloth than can be woven on one loom. Whilst they can be mounted on the loom they are, because of their cost, more frequently mounted on a separate cutting or inspection table.

**Leno and Helical Selvedges**

In most shuttleless weaving machines a length of weft has to be cut for every pick. For looms not fitted with tucking motions the warp threads at the edges of the fabric have to be locked into position to prevent fraying. With some weft insertion systems the weft has to be cut only after every second pick and it is possible to have a hairpin selvedge on one side of the fabric and a locked selvedge on the second side. Leno and helical selvedges are widely used to lock warp yarns into position respectively. When shuttleless weaving machines were first introduced there was considerable customer resistance to fringe selvedges but, in the meantime, it has been found that they meet most requirements.

With leno selvedges a set of threads at the edge of the fabric is interlaced with a gauze weave which locks round the weft thread and prevents ravelling of the warp. As the precut length of the picks always varies slightly, a dummy selvedge is sometimes used at the edge of the cloth. This makes it possible to cut the weft close to the body of the cloth resulting in a narrow fringe which has a better appearance than the selvedge from which weft threads of varying lengths protrude. It also has the advantage in finishing and coating that there are no long lengths of loose weft that can untwist and shed fly. Because of the tails of weft and because of the warp in dummy selvedges, more waste is generally made in narrow shuttleless woven fabrics than in fabrics woven on shuttle looms. For wide cloth the reverse frequently applies because there is no shuttle waste. Leno selvedges, sometimes referred to as 'centre selvedges', may be introduced into the body of the cloth if it is intended to slit the width of cloth produced on the loom into two or more widths either on the loom or after finishing.

Helical selvedges consist of a set of threads which make a half or complete revolution around one another between picks. They can be used instead of leno selvedges and tend to have a neater appearance.

## THE DEVELOPMENT OF FRAMES AND TREADLES

During the Shang Period (1766 to 1122 BC) in China the treadle and frame loom system was developed. Weaving frames are rectangular shaped structures that hold the heddles. The heddle is a long needle-like string or metal device with an opening (called an eye) in the middle. There is a loop (called a doup) at the top and bottom of each heddle.

The top and bottom doups are threaded on to the top and bottom cross bars of the weaving frame. The warp (vertical) yarns are threaded through the eye of the heddle.

A treadle is a foot- or hand-operated lever that is tied by a cord to the bottom of the weaving frame. Using a system of ropes and rollers the harnesses are moved up and down. This system produces an opening in the warp threads called a shed. The weaver passes the weft (horizontal) threads through the warp shed to weave the cloth.

## THE CONSEQUENCES OF FRAMES AND TREADLES

The development of the treadle and frame system freed the weaver of the task of picking each warp (vertical) thread by hand. As a result, a weaver could produce large quantities of cloth in a relatively short time.

The early treadle and frame loom system is the basis of modern shed looms widely used by both hand weavers and industrial weavers. [SB] {WT}

## THE MIDDLE AGES

Cloth weaving during the middle ages developed swiftly. Weavers developed many clever changes to the original frame looms and shed loom systems.

By the 11th century many of the weaving patterns used today had been invented. Skilled weavers developed highly specialized cloth.

During this time the task of weaving cloth began slowly to move away from the family unit into specialized work places. [SB] {WW}

## THE DRAW LOOM

By 1400, the draw loom was widely used. The Draw Loom is a very large and complicated loom.

The early Draw Looms required 2 people, the weaver and the draw boy, to operate them. The draw boy sits on top of the loom. Under the direction of the weaver he raises and lowers individual heddles attached to each pattern warp thread. This method of weaving produces cloth with very intricate patterns. [SB] {MW}

## GUILD STRUCTURE

During the Middle Ages the Craft Guild system developed. A weaving guild was an organized group of people involved in producing cloth. The Weaving Guild was made up of guild workshops. [SB] {SW}

## GUILD WORKSHOPS

Guild workshops were home-based businesses where cloth was produced.

## GUILD APPRENTICES AND JOURNEYMEN

In the Guild Workshop the Master weaver directs the work of the apprentices and journeymen. Apprentices are people learning to be weavers. The journeymen are weavers who have finished there training as weavers but continue to work in the Master's Shop. [SB] {SW}

## THE ROLE OF THE GUILD

The Guild demanded high standards and excellence in cloth production. Also, the Guild severely limited the number of weavers who could become Masters. In this way the Guild system was able to set prices and control the business practices of weavers.

The Guild Workshops of the Middle Ages were the foundation for textile production in the Industrial Revolution. [SB] {SW}

## THE INDUSTRIAL REVOLUTION (1760-1815)

Cloth weaving became a mechanized industry with the development of steam and water powered looms. The invention of the fly shuttle removed the need to have a weaver place the weft (vertical) thread into the warp (horizontal threads) by hand.

## THE FLY SHUTTLE

A fly shuttle is a mechanical device using ropes and pulleys to deliver the weft (horizontal) thread into the warp (vertical threads). The weft yarn is wound on to a bobbin and it is placed in a fly shuttle. A fly shuttle is a long, narrow canoe-shaped instrument, usually made of wood, which holds the bobbin.

After the shuttle is loaded with the weft-filled bobbin, the weaver places the shuttle onto the *shuttle race*, a small, narrow shelf that the shuttle glides along as it goes in and out of the weft. The weaver pulls on a rope attached to the fly shuttle mechanism and this propels the shuttle across the weft. The invention of the fly shuttle increased the volume of cloth production and forced technological advancement in the spinning industry to supply larger amount of yarns.

### THE JACQUARD MACHINE

The early 1800s saw the development of the Jacquard Machine. This revolutionary machine used a punch card mechanism to operate the loom and is credited as the basis of modern computer science.

This complicated machine was added to the top of the weaving loom. A series of card with holes punched in them is continuously run through the machine. The Jacquard Machine is able to move individual warp threads up and down according to the pattern of holes punched into the cards.

Cloth woven on a loom with a Jacquard Machine can have very intricate patterns.

### INDUSTRIAL CLOTH PRODUCTION

The technological innovations in cloth production made during the Industrial Revolution dramatically changed the role of the weaver. Large volumes of inexpensive cloth were now readily available. Weaving had been changed to a manufacturing industry. Textile workers were among the founders of the modern labour movements.

### WEAVING TODAY

Today most of our textile needs are supplied by commercially woven cloth. A large and complex cloth making industry uses automated machines to produce our textiles.

However, there are artisans making cloth on hand looms, in home studios or small weaving businesses, who keep alive the skills and traditions of the early weavers.

## EARLIER SPINNING AND WEAVING

Weaving. When or where man first began to weave cloth is not known, nor is it known whether this art sprang from one common centre or was invented by many who dwelt in different parts of the world. There is such a sameness in the early devices for spinning and weaving that among some men of science it is thought that the art must have come from a common centre.

Fabrics were made on the farms two or three hundred years ago in the following manner: the men of the household raised the flocks, while the women spun the yarn and wove the fabrics.

In this way the industry prospered, giving occupation and income to thousands of the agricultural class. You might say that in England fabrics were a by-product of agriculture. As time went on, farmers of certain sections of England became more expert in the art, and the weaving became separated from the spinning. The weavers became clustered in certain towns on account of the higher skill required for the finer fabrics. The rough work of farming made the hands of the weaver less skilful. This, coupled with the fact that the looms became more complicated with improvements, called for a more experienced man. Great inventions brought about a more rapid development of the factory.

Richard Arkwright, who has been called the "father of the factory system," built the first cotton mill in the world in Nottingham in 1769. The wheels were turned by horses.

In 1771 Arkwright erected at Crawford a new mill which was turned by water power and supplied with machinery to accomplish the whole operation of cotton spinning in one mill, the first machine receiving the cotton as it came from the bale and the last winding the cotton yarn upon the bobbins. Children were employed in this mill, as they were found to be more dexterous in tying the broken ends. As the result of this great invention, factories sprang up everywhere in England, changing the country scene into a collection of factories, with tall chimneys, brick buildings, and streets.

From 1730 to the middle of the nineteenth century the development of inventions was rapid:

- 1730—First cotton yarn spun in England by machinery by Wyatt.
- 1733—English patent granted John Kay for the invention of the fly shuttle.
- 1738—Patent granted Lewis Paul for the spinning machinery supposed to have been invented by Wyatt.
- 1742—First mill for spinning cotton built at Birmingham; moved by asses; but not successful.
- 1748—Patent on a cylinder card as first used by hand, granted Lewis Paul.
- 1750—Fly shuttle in general use in England.
- 1756—Cotton velvets and quiltings first made in England.
- 1760—Stock cards first used for cotton by J. Hargreave. Drop box invented by Kay.
- 1762-67—Spinning-jenny invented by Hargreave.
- 1769—Arkwright obtains his first patent on spinning.
- 1774—Bill passed in England to prevent the export of cotton machinery.
- 1775—Second patent of Arkwright on carding, drawing, and spinning.
- 1779—Mule spinning invented by Crompton. Peele's patent on carding, roving, and spinning.
- 1782—Date of Watt's patent for the steam-engine.

- 1783—Bounty granted in England for the export of certain cotton goods.
- 1785—Power loom invented by Cartwright. Cylinder printing invented by Bell. A warp stop-motion described in Cartwright's patent.
- 1788—First cotton factory built in the United States, at Beverly.
- 1789—Sea Island cotton first planted in the United States. Samuel Slater starts cotton machinery in New York.
- 1790—First cotton factory built in Rhode Island by Slater.
- 1792—First American loom patent granted to Kirk and Leslie.
- 1794—Cotton-gin patented by Eli Whitney.
- 1801—Date given for invention of the Jacquard machine in France.
- 1803—Dressing machine and warper invented in England by Radcliffe, Ross, and Johnson.
- 1804—First cotton mill built in New Hampshire, at New Ipswich.
- 1805—Power loom successfully introduced in England after many failures.
- 1806—First cotton mill built in Connecticut, at Pomfret.
- 1809—First cotton mill built in Maine, at Brunswick.
- 1812—First cotton mill built at Fall River.
- 1814—Cotton opener with lap attachment invented in England by Creighton.
- 1815—Power loom introduced into the United States at Waltham.
- 1816—First loom temple of Ira Draper patented in the United States.
- 1818—Machinery for preparing sewing cotton invented in England by Holt.
- 1822—First cotton factory erected at Lowell.
- 1823—Differential motion for roving frames patented by Arnold. First export of raw cotton from Egypt to England.
- 1824—Tube frame or speeder patented by Danforth.
- 1825—Self-acting mule patented in England by Roberts.
- 1828—Ring spinning patented by John Thorpe. Cap spinning patented by Danforth.
- 1829—Revolving loom temple improvements patented by Ira Draper.
- 1832—Stop-motion for drawing frames invented by Bachelder.
- 1833—Ring spinning frames first built by William Mason.
- 1834—Weft fork patented in England by Ramsbottom and Hope. Shuttle-changing loom by Reid and Johnson.
- 1840—Automatic loom led off. Important temple improvement.
- 1849—First cotton mill erected in Lawrence.

Through this great change from hand to power work, thousands were thrown out of employment in the great textile centres, and much suffering occurred, which led to the smashing of machinery.

## KNITTING MACHINERY

Like many other industries, the hosiery trade owes its first and most important impetus to the genius of one who was not connected with the business in a practical way. This event took place when the Rev. William Lee invented the hand frame. He was married early in life, and his wife was obliged, on account of the slender family finances, to knit continuously at home. Struck with the monotony and toil involved in knitting with the hand pins, Mr. Lee evolved a means of knitting by machinery and brought out the hand stocking-frame, which today preserves its chief features very much as Lee invented them.

When knitting by hand, one must form each loop separately, and loop follows loop laboriously until the width of fabric has been worked. Lee contrived to make the whole row of loops across the width simultaneously by arranging a needle for each loop and placing in connection with each needle a sinker and other apparatus for completing the formation of the loop. First of all, the yarn is laid over the needles, which are arranged horizontally, and the sinkers come down on the yarn and cause it to form partial loops between the needles. The old loops of the previous course are now brought forward and the new yarn is drawn through them in the same way as is done on the hand pins. Thus the new yarn of one course is drawn through the loops of the preceding one, and so the whole fabric is built up. This frame of Lee's held its own in the great centres until some thirty years ago.

Lee's hand frame gave way to what is termed the jack and sinker rotary frame, which was like the hand frame in its chief features, but with the advantage that all the motions were brought about by power. The various operations were put under the control of a set of cams and made to perform their movements in exactly the same way as in the case of the hand frame. In the first power machine for knitting, the machine builder used the cam mechanism, and in examining the latest machines we find that he has persisted in this course throughout.

The cam movement is characterized by great smoothness of working and absence of vibration, which is very necessary in a machine of the delicate adjustment of the knitting frame. It is usual to connect some of the parts with two of these cams, one of which controls the up-and-down motion and the other the out-and-in movement. When these two cams work in conjunction, we obtain all the possible degrees of harmonic motion.

From the jack and sinker frame the next really important step was taken when William Cotton brought out his famous Cotton's patent frame. In his machine the frame was in a sense turned on its back, for the parts, such as the needles, which had been horizontal, were made vertical and vice versa. He also reduced the number of the moving parts and perfected the cam arrangement. Another very important development of the machine was when it was built in a number of divisions so as to work a number of articles side by side at one time. At present there are knitting frames which can make twelve full-sized garments at one and the same time. Another important improvement was effected when the fashioning apparatus was supplied to the machine, by means of which the garments could be shaped according to the human form by increasing or decreasing the width as desired.

## METHODS OF SPLICING MATERIALS FOR WEAVING

CUCH materials as carpet and oriental wools, fine worsteds, carpet ravelings, darning and knitting cotton should, in splicing, be run past each other. In weaving, run the wool through the warp to the very end. Start the new piece a few warp threads back, being careful to go over and under exactly the *same* warp threads as you did when finishing the end. As you pass these threads you will find that you are taking tip the right warp threads, and that no mistake has been made. It is best to run the threads past each other in the *middle* of the mat rather than on the. *sides.* The children learn this method of splicing very quickly and the result is much more satisfactory than knotting, because the back of the rug or mat will be smooth. As Mrs. Wiggin says: "There should never be a wrong side to work any more than there should be to folks."

In splicing such materials as silkoline, rags, candle-wicking, chenille, and macreme cord, lay the end of one piece over another, each lapping about one-quarter inch, and sew securely with silk or thread of like colour. Cut off the selvedge ends of rags. These strips can be ran past each other, but the work will not be so smooth. In splicing Germantown wool, heavy worsteds, or rope silk, thread a worsted needle with one strand obtained by unwinding the wool or silk, lay one end over the other, and sew over and over. Twist the part just sewn between the thumb and finger and the splicing will be hardly visible.

When weaving stripes, splice the wool so that the piecing will come on top of the rod. In this way the new colour will start at the edge of the rug, as it should, and the number of loops on the rod will be the same on each side. Consider the *under* side of the weaving as the *right* side. It is always smoother and cleaner, and the splicing can be done more neatly on top of the rod.

Splicing raffia is the most difficult of all, and the method used in braiding and basket weaving is the best. As you near the end of a strip in weaving it usually becomes narrower. Find another strip having a narrow end, and place one over the other, securing, if necessary, by winding a very narrow piece—just a thread torn from a long piece — and fastening this by sewing a few times over and over. Or, the two narrow ends may be run past each other, as in carpet ravelings. Care should be taken to have the splicing the same width as the other parts of the weaving, so that the spliced parts will not be noticeable. Leather, leatherette, and celluloid strips should be long enough to extend the entire width and length of the frame without splicing. The ends can be cut, as is done in paper weaving, or turned in some pretty way like that in the splint work.

## METHODS OF STRINGING WARP

THE adjustable loom can be strung with warp of three widths,, one-half inch, three-eighths inch, and three-sixteenths inch, thus giving opportunity for a variety of materials. For heavy rags, candle wicking, etc., wind the warp strings around three teeth in the head and foot pieces. This will give a warp of one-half inch —that is, one-half inch from one string to the other. For silk, silkoline, finer rags, carpet ravel-ings, double wool, etc., wind the warp strings around two teeth, thus making a warp of three-eighths inches.

For double wool, worsted, rope silk, chenille, or raffia, where one wishes to reproduce kindergarten designs, as in paper-weaving, place the warp strings around one tooth only. This makes a close warp of three-sixteenths inch, which helps to form the design with the woof threads. In this case the warp should be of the same material as the woof. In kindergarten patterns the woof threads determine the colour effect. It is better to have the children weave the pattern first with practice mats and slats, particularly if they have never had experience in the kindergarten. Suggestions for weaving kindergarten designs are given under the head of *Raffia.* For a plaid effect, string the warp at regular intervals, with different colours. Then weave the same coloi-s at equal intervals to form the plaid. Shawls, carriage blankets, etc., woven in this way are very attractive. A striped warp is strung in the same way. The stripes could be continued through the mat, if desired, by weaving only *one* colour in the woof. By weaving *two* colours squares are obtained such as those seen in the corners. For weaving with carpet ravelings or rags, and sometimes double wool, where a plain effect is desired, the warp should be of common twine, as near the colour of the work as possible. Carpet thread is good, especially for the double warp in Turkish rugs. Balls of warp string can be obtained at department stores. Oriental cord comes in several colours, and can be had at a few cents a ball at the notion and stationery counters in department stores.

The warp should always be one continuous string, and several inches should be left at each end in order to fasten securely when the work is finished. If preferred, the warp ends can be fastened before the weaving is commenced. Care should be taken to place the first and last strings of the warp directly *over* the rods, and, in weaving, to pass the woof threads entirely around the *rods and strings* to insure straight edges.

The ends of wool warp threads should be wound in and out of the notches to the right and left of rods, to fasten them until the wcavino- is finished. It sometimes happens that little children, and more especially those who are blind, pull up the warp strings when near the end of the work. In such cases it is a good plan to pass a rubber band *over* the warp strings at the top of the loom and *behind* the bars, back of the head piece, making it set up close by putting it around one tooth at each end. In this way the warp strings cannot possibly slip out of the notches.

Some teachers splice the warp with a weaver's knot, an illustration of which can be seen in any large dictionary. The continuous string is to be preferred, however, as experience has proved that even a weaver's knot will sometimes fail to stand the stress of weaving. It is very difficult to splice a warp of raffia. It is better to knot the warp threads in pairs, leaving two or three inches beyond the head and foot. These ends may be used for a fringe by tearing very fine, or they may be run down in the woven part with a darning needle, as rattan is run down in basket work.

When the weaving is done and the mat lifted from the loom, the ends of the wool warp strings can be run in along the sides with a tape needle. If the warp be of twine, it is better to tie the end to the next warp string and allow the fringe to cover the knot; or, as in the case of silkolinc.thc woof strips can be caught over the warp strings with silk of the same colour in order to hide them. Only experience can teach the. tightness with which a warp should be strung. Worsted, carpet thread and twine will stretch as the work progresses, and raffia will not. If the warp be too loose the work will be uneven and the strings will slip out of the notches. If it be too tight it will be difficult to finish the last two or three inches and the woof threads will look crowded. The best test is to place the hand upon the warp before commencing to weave. If it feels firm and does not push down too easily, but springs slightly beneath the hand, it is probably correct.

Where the warp is of the same material as the woof and it is desired to extend it to form a fringe, it can be done in the following manner.- After the loom is adjusted for the size required, cut the warp strings so as to allow two or three inches beyond the head and foot pieces. If you intend to knot the fringe in some fanciful way after the weaving is finished, allow four or five inches. Take two threads, knot so as to leave the required length for fringe below the foot piece, then pass around one or two teeth, a.s the case may be, draw tightly to the head piece and knot firmly on the upper side, leaving a fringe of the same length there. Knot the strings in pairs in this way until the whole warp is strung.

It will be noticed that the rods are placed beneath the notches of odd numbers. In knotting warp strings in pairs it will be found necessary, when the last tooth is reached, to do one of two things — either allow one string to lie beyond the rod, or, having strung the warp within one tooth of the rod, to start the next string in the *same* notch, bringing the two strings together. This will bring one string on top of the rod and none beyond. In the first case, the string beyond the rod must be taken up in weaving with the one on top of the rod. Experience has proved the second method to be the better one. I Kiz-Kilim rugs have perforated or openwork patterns. To produce this effect string a *double* warp through every notch in the foot and head pieces; that is, use two strings in each notch, tying in pairs for fringe as before. Use a brownish white carpet thread. With strong black thread string through every other notch to outline perpendicular sides of squares in the pattern. Your warp will be strung three-sixteenths inch, but the black threads will be three-eighths inch. This will enable you to keep the patterns straight as the work progresses.

In drawing designs for Turkish rugs, where the pattern is to be placed under the warp, it is better to make a squared paper first. Lay the head piece of the loom upon unlined paper. Place a dot at every other notch. Draw perpendicular lines first, then dot for horizontal lines. The result will be a foundation to fit your loom. If the squared paper of the kindergarten be used the squares will be either too large or too small to correspond with the notches in the loom. It will be found very easy to transfer a pattern from a rug to the paper. Fasten the pattern under the warp by overhanding to the rods, taking care to have the black strings directly over the perpendicular lines in the pattern.

## STEPS IN WEAVINC

'PHE principles of weaving are very easily learned with felt mats and slats. One-half a yard of felt two yards wide will make thirty-six mats six inches square. These are very durable, and can be used year aftcr year, if protected from moth during the summer. Some prefer leather or oil-cloth mats, backed with heavy unbleached muslin, but they are more expensive, and not so pleasant to work with as the soft wool. The slats, which should be at least one-half an inch wide, can be obtained at any kindergarten supply store. Buy the uncolored slats and dye them yourself. Dark green mats, woven with deep red slats, are pretty. The slats are easier to handle if they are soaked and cut the required length before dyeing.

When the six-inch mats are cut, allow a three-quarter-inch margin on all sides. Measure the mat for one-half-inch strips, of which there will be nine, and mark by snapping a chalked string upon the mat. Double it with chalked lines outside and commence to cut from the centre; then open and finish cutting to the margin. It would be better for very little children if the strips and slats could be one inch wide. In this case the mats would, of course, be larger, and it might be necessary to have the slats made to order. The slats should be kept in little bundles containing the required number, and secured by rubber bands. If one could have plenty of time and material it would be a good plan to have several sets of mats of different sizes, so that the children would not always be confined to one number and its combinations in a certain set of patterns — in this case, nine — but have the pleasure which conies from variety. Demonstration cards and diagrams for weaving can be obtained at the kindergarten and school-supply stores. An illustration of an excellent demonstration frame can be seen in the " Kindergarten Guide," by Lois Bates. Sample mats can be woven by the older children from the designs in any of the "Guides," and given to the smaller children to copy.

When the purpose of these practice mats is understood there can be no objection to them on the ground that the work is destroyed by pulling out the slats each time. It is not an unusual thing to see in schools, and even in kindergartens, faithful and conscientious teachers remaining after hours to pull out the slats, on the principle, perhaps, that what the children do not see will not affect their development, and the innocent little bundles are given out again on the morrow, only to undergo the same experience at night. One wonders sometimes if this is possibly within the definition of deception. " We mount to the summit, round by round," and when the children understand that in doing the work with the slats well, they are only learning *how,* and that each successful attempt brings the delightful day nearer when they may have a loom to work upon, they are perfectly satisfied.

When the children have learned to weave the small mats, further practice can be had by weaving long slats into a warp of cord on the loom. It is better to conquer the mystery of " over and under" in this way than to undo the work and wear out the material after making a mistake. Many teachers prefer to make the practice mats of paper because they are cheaper. Heavy paper, in desirable colours, can be obtained at the wholesale paper houses, and for a small sum can be cut in squares of any required size. Mats can be made more durable by pasting them on heavy muslin before cutting. In many schools children in grades above the entering room prepare their own mats by measuring with tablets or rulers and then drawing and cutting on the lines. When they have learned to do them well, let each child make one for the entering room. Nothing strengthens the community feeling so much in a school as to encourage the older pupils to help the younger.

The mat-weaving, as it is done in the kin- *Mat*-dergarten, is very beautiful and fascinating *the* work. The mats can be obtained in any size *trarten* and any width of strips at the supply stores. The weaving is done with a long steel needle which has a spring at one end to hold the strip. After preliminary work with the felt mats and slats the children find themselves able to weave quite independently, particularly if demonstration cards or sample mats are placed before the class. An infinite variety of patterns, which later will be useful in wool-weaving, can be found in the " Kindergarten Guides." In weaving patterns having a centre, it is better to weave two strips at once, pushing one to the top and one to the bottom of the mat. The old numbers of the Godey and Peterson magazines have patterns for Berlin wool and bead work which can be used for the paper mats with good effect. Mrs. Kate Douglas Wiggin (Mrs. Riggs) has some good suggestions for invention in weaving, in her " Republic of Childhood" (Occupations). The value of weaving in number work is also admirably set forth in this book.

At Christmas time many charming little gifts can be made of these mats. Sachet cases made of a six or eight inch square, with four corners folded to the centre, are attract ive. Inclose a square of wadding, in which a pinch of heliotrope or white rose perfume powder has been hidden, and fasten the corners together with a scrap picture of old Santa Claus. Slat work is useful in learning the fundamental principles of weaving, although this work is more closely related to basket than to rug-weaving. It is an excellent preparation for the free-paper weaving, and is also a step toward basket work.

In interlacing slats the mystery of " over and under " is solved and the dependence of one slat upon another in making a perfect whole is shown in a forcible way, particularly when the form falls to pieces in the attempt to lift it from the table. Edward Wiebe says in his "Paradise of Childhood": "It was the *one* slat which, owing to its dereliction in performing its duty, destroyed the figure and prevented all the other slats from performing theirs." One experience of this kind will teach more than a thousand precepts. The geometrical forms learned in the sense-training lessons can be reproduced with the slats and will thus be impressed upon the mind during the period of busy work at the desk. A series of beautiful designs is published by E. Steiger, New York. Many designs may be grouped for decoration, and single symmetrical figures can be mounted upon heavy paper.

Free-paper weaving requires quite a little skill of hand and a great deal of patience before the child can achieve a successful result. Perhaps a few words regarding it, and information about a simple sequence of paper patterns, will not be out of place, since so many are today taking it up. Strips of manilla paper forty inches long and one inch wide are used. These are cut into strips eight inches, sixteen inches, twenty inches, and twenty-four inches in length. For the first pattern of the sequence take four strips eight inches long and double each one. Hold two of them side by side in the left hand, so that the open ends of the outer strip are at the top while those of the other are at the bottom. With the right hand inclose the first strip in the left hand with one of the remaining double strips and pass the ends of the latter between the two ends of the second strip. Then hold the work in the light hand and proceed in the same way with the left hand. When both strips are in, draw them tight and they will be firmly woven. The ends can be cut in any way desired. These little forms can be used for bookmarks. They are very attractive when made in two tones of one colour.

The second pattern of the sequence is made with sixteen-inch strips. The first part is woven like the bookmark. Four double strips now project from the square. Begin at the bottom and fold back the *upper* one of each of these double strips. As you do this you will find that you are weaving another square on top of the first one.

To secure the last strip pass it under the square next to it and pull it through. You will now have eight single strips, two on each side. To form these into points for a star proceed as follows: Begin with the right-hand strip at the top and number all the strips from one to eight. Fold number one back toward the right, making at the fold a right-angled triangle. Fold the strip down again towards you, making another triangle which is folded back to the left on the first one. Slip the end of the strip under the square next to it and cut it off. Proceed in the same way wi th three, five, and seven. Then turn the form over and fold the strips two, four, six, and eight in the same way, cutting off the strips when finished. Many of these stars can be joined to make mats, baskets, picture frames, etc. They are pretty when made of gilt or coloured paper for Christmas decorations.

Pattern number three, a bookmark, is made like the first, except that eight strips of sixteen-inch length are used and the strips woven at right and left are finished as directed for the mat. Number four is another form like this, with the long, ends back and front slipped through squares to form a napkin ring. Number five is a six-inch mat made of twelve twenty-inch strips. Weave six double strips left and right into two strips and then add four to make the square. To finish the edge cut off the *under* one of each double strip, fold the upper one over it and then slip it under the square which comes next, cutting it off even. Strips of felt can be woven in this way for table mats or holders.

# 7

# Textile and Handicraft

Handicraft which is popularly known as artisanal handicraft or artisan's is creation. It is a main sector of traditional craft and usually the term is applied to traditional means of creating items. Such items often have cultural or religious significance. The items made by ma ss production or machines are not handicraft goods. Handicrafts require those skills which facilitate decorative things or articles to be made completely by hand, using some basic tools. Some crafts are passed on from generation to generation, such as dyeing and printing.

The history of printing started around 3000 BC with the duplication of images. The use of round "cylinder seals" for rolling an impression onto clay Tablets goes back to early Mesopotamian civilization before 3000 BC, which are the most common works of art to survive and feature complex and beautiful images. In both China and Egypt, use of small stamps for seals preceded the use of larger blocks (Wikipedia, 2013).

Printing on fabric has a long history, with archaeological evidence from ancient India, Egypt and Greece. The lines of development from the earliest printed textiles through to modern methods involve science, technology, politics, exploration and trade.

Textile printing is the process of applying colours to the fabric in definite patterns or designs. In printing, wooden blocks, stencils, engraved plates, rollers or screens can be used to apply colours on the fabric. Colourants used in printing basically contain dyes, which are thickened to prevent the colour from spreading on the fabric by capillary action beyond the limits of the pattern or design to be printed.

Garment printing or textile printing is not just done for adding value to the clothes but also to make them more appealing and attractive to the eye. When talking about printing, the whole cloth is printed either in bold or in small printing designs which can be done by hands as well. Knowing the importance of printing and keeping in mind its attributes like colour fastness, neatness etc.

Textile printing is also related to dyeing but the basic difference in both is that in the case of dyeing whole fabric is uniformly covered with one colour, while in printing one or more colours are applied to certain parts only and in sharply defined patterns. Printing is also called as localized dyeing. Block printing can be categorised into two types i.e. hand block printing and machine block printing.

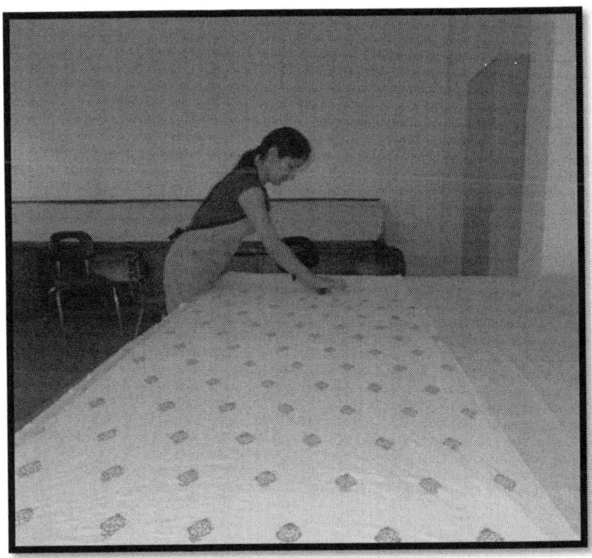

**Fig.** Traditional hand block printing

In India, the art of hand block printing has passed from generation to generation, and has traditionally been done by using natural dyes. Various fabrics/ garments like sarees, kurtas, shirts, salwar kameez, dupattas, skirts etc are made using block printing. The process of hand block printing includes artisans to soak carved wooden blocks in different colors and then paste them on the fabric thus creating some magical wonder on the piece of cloth. It is a complex and labour- intensive craft that involves a variety of skills at different stages like carving the block (usually done by craftsmen), preparing the cloth, mixing dyestuffs, and finally the printing, dyeing, and washing steps, which may be repeated several times to obtain a final desired colored and designed fabric.

Blocks used for block printing are usually made out of wood, which are hand carved by skilled crafts men. The blocks called bunta comes in various designs and sizes.

The underside of the block has a design engraved on it. Each block has a wooden handle and two to three cylindrical holes drilled into the block for free air passage and also to allow release of excess printing paste. A separate block is used for each colour and pitch pins at the corners guides placing of the blocks to assure accurate repeat of the pattern. Sometimes 5-8 blocks are used depending on the design and the cost of block making and production goes up accordingly.

**Fig.** (a) Cylindrical holes drilled into the blocks (b) Design engraved underside of the block (c) Wooden handle attached on the back side of the blocks

The typical hand block printing had no large, uniform areas of colour but was skillfully built up from many small colored areas, because wooden surfaces larger than about 10 mm in width would not give an even print. This had the advantage that a motif such as a flower would have an effect of light and shade obtained from three or four blocks, each printing has a different depth of the same colour or a different hue. This perceptibly meant that a number of blocks were required, and considerable care was needed in fitting the adjacent parts of the design.

**Fig.** Resist printing style

Block printing allows for the maximum play of imagination and creativity on cloth. One can change the block's direction, experiment with the placement of each block and change colours.

Three main tools of a block printed fabric are wooden blocks, fabric and dye. In properly printed fabrics the colour is bonded with the fibre, so as to resist removal due to washing or friction.

Traditional hand block printing techniques may be broadly categorised into three styles: Direct printing. It can be done on white or a coloured fabric. If done on coloured fabric, it is known as overprinting.

Resist printing, is a style in which the part of the cloth which is not to be dyed is covered with the paste of resin and clay. Then the fabric is dyed with desirable colour and at this stage the dye penetrates through cracks which create wavy effect of colors on the cloth.

Discharge printing is different style of printing in which a bleaching agent is printed onto previously dyed fabrics to remove some or the entire colour, so that a printed design is created. In this type of printing each step contributes to fixing the dyes and making the colours rich and vibrant.

Today, India is a major centre of hand block printing and states of Madhya Pradesh, Rajasthan and Gujarat are the flourishing trade centers for this printing. The Ajrak resist-printing technique is found in Anjar and Dhamadka in Kutch (Gujrat). Javad prints in Indigo and Alizarine are mostly used in Madhya Pradesh. Bagh, which lends its name to the *bagh* prints, is also famous in Madhya Pradesh. In states like Gujarat and Rajasthan, two styles of hand block printing are particularly employed namely Sanganeri and Bagru. The main difference between the two is the background on which printing is done.

Rajasthan and Surat in Gujarat have become the important trading centers of printed textiles particularly in block print art. Block printing has earned a reputation for itself in Rajasthan for ages. It has a long inheritance of its fine hand block printed textiles. The craft of printing is practised since many previous centuries by skilled artisans i.e. dabu (resist-printing); bagru, pigment, discharge, embossed printing etc.

A stability of tradition is visible in the similar design styles of printing which is found even today.

**Fig.** Discharge printing style

The craft of block printing is practised in almost every village in Rajasthan. As it is heavily reliant on water sources; hence initially commercial printing centres began to rise near water sources, the most famous centres being Sanganer and Bagru near Jaipur, Barmer, Jodhpur and Akola near Udaipur.

Over the time, each centre for block printing in Rajasthan has developed its distinguished design style and techniques.

According to tradition in Rajasthan, craft skills are passed down the generations, from parent to child, the expertise remains within the family and people engaged in this, printing trade form an identifiable group called the '*Chhipa community*'.

Rajasthan is the heartland of hand block printing. Sanganer is virtually flooded with block - makers and printers to the extent that textiles are hand - block printed in the courtyards of homes. These prints command a huge demand from major fashion centres world - wide.

The distinctive feature of Sanganeri print is that its background is always white. Black colour is used for outlines whereas red colour is used to fill in the figures and floral motifs.

The chief colour used is orange and red with floral prints in yellow and blue - black. The art of Khari or overprinting in gold is also practised here.

Bagru is a small village located on Ajmer road and approximately 32 kilometers far from centre of Pink-City (Jaipur) which is famous for its hand block printing.

The prints which are popularly known as 'bagru prints' make the village a crafts centre well known in Jaipur and in surrounding areas.

Apart from this, Bagru craft is also famous for its ecological consciousness and the use of traditional dyes. This tradition of hand block printing, which is more than three hundred years old, has still been kept alive by the efforts of the Bagru artisans. Bagru printing is a textile design in which vegetable colours are used it is the specialty of Rajasthan block printing.

Still today, the craftsmen following the original tradition to wipe the cloth with fuller's earth and immerse it in turmeric water to provide the cloth with the customary cream colour. Following this, various designs are embossed on the fabric, with the help of natural dyes of earthly shades. They use simple and eco-friendly tools along with natural and vegetable dyes to print the cloth.

The craftsmen make blue colour from indigo, red colour from madder root, green from indigo mixed with pomegranate juice and yellow from turmeric. Excellent bed covers and other materials are manufactured by these villagers.

Another technique employed was printing with the use of mordants. Mordants are chemicals or natural sources that absorb the dye. The cloth was firstly printed with mordants and then immersed in a dye bath.

Only the sections that had absorbed the mordant absorb the dye. The cloth was then washed in flowing water and spread out to dry on the riverbank allowing sun to develop the colour. Then the untreated sections were bleached with local ingredients like goat droppings etc.

The 'lake city' Udaipur has always been a centre for embossed effect printing. Today, very few families are doing this work and the designs used are mostly traditional in nature. These special embossed effects are obtained by Khari printing in which gold or silver powder is mixed with various other ingredients and painted directly on to the cloth.

**Fig.** Embossed or Khari printing

Akola is one of the villages, in Bhopalsagar Mandal in Chittorgarh District of Rajasthan state. It is 88 kilometres far from its district main city Chittorgarh and situated at the bank of Bedhach River. This village is famous for resist block printing on fabric, this printing is known as Akola printing, Dabu printing or mud resist printing. Dabu is the local name for the resist print process in which mud as resist paste is applied by hand, using wooden block to seal the fabric form the effect of further treatment. The 95 families involved in this profession are known as *"Chippas"* so this village is also known as *"Chippon ka Akola"*.

**Fig.** Dabu mud resist printing

In block printing colour variations is not an easy task. The colors are generally tested for their fastness and effect before applying on the garment. Different types of dyes are used for printing cotton fabric- such as indigo, pigment dyes, rapid fast dyes and vegetable dyes.

The rapid dyes once prepared for printing have to be utilized on that day only. The pigments, which are not truly dyes, are also used extensively for printing.

Pigment colors are used more because this method is simpler as compared to other dyes and after mixing them for printing it is not necessary to use in the same day, it can be stored for some time. The pigment colors before the application has to mix in right proportion with binder and fixer that are very resistant to laundering or dry-cleaning. Pigments are among the fastest known colours and are effective for light to medium shades.

**Fig.** Rapid printing

Discharge printing also called as extract printing. It is a method of applying a design to dyed fabric by printing a colour-removing agent, such as chlorine or hydrosulphite, to bleach out a white or light pattern on the darker coloured ground. In colour-discharge printing, a dye not influenced from the bleaching agent is combined with it, producing a coloured design instead of white on the dyed ground.

**Fig.** Discharge printing

In earlier times, the royal family of Jaipur was the chief patron of the craft. After independence, the craft almost died till it was revived in the 70s owing to the patronage of prominent exponents like Kamala Devi Chattopadhyay, the person behind the cottage industry movement, Prabha Shah, Laila Tyabji, Pupul Jayakar and John Singh of Anokhi. But faces still look grim. Faced with pollution charges and that too, only because of the screen printing industry, life is still dark. The good news is that, because of the constant efforts, Under Geographical Indication of Goods Registration and Protection Act.1999, Sanganeri Hand Block Textiles and Furnishings' have been recognized to have a unique identity in manufacturing and belonging to the area of Sanganer. Consequently sale of hand block or screen printed product from any other area as of Sanganeri will be punishable under the law, thereby giving a new lease of life to the trade and some 1500 artisans and about 5000 families indirectly or directly dependent on block printing (AIEED, 2013).

The artisans try hard on developing the art completely through natural means using vegetable dyes and natural colors for printing and preserving their workmanship by passing it over to the coming generations. Hand printing of clothes and textiles has been practiced since the ancient times in Rajasthan. These prints reflect the history, tradition and lifestyle of the native people. In the present context, these prints assume significance. As the world is waking up to environmental consciousness and eco-living, the ancient art and craft form becomes special because it is "high on ecological consciousness and uses eco-friendly hand block printing practices."

The art of making natural dyes is one of the oldest known to human. In India, it was used for colouring fabric and other materials. Use of mordanted cotton yarn and fabric found in Indus valley site of Mohenjo-Daro. The Indian craftsmen and their knowledge of the technology of mordants and its use in textiles remained protected for a long time. India had a virtual monopoly in the production of dyed, painted and printed textiles. Use of non-allergic, non-toxic and ecofriendly natural dyes on textiles has become a matter of significant importance due to the increased environmental awareness in order to avoid some hazardous synthetic dyes. Very few natural dyes are colour fast with fabric. Mordants are substances which are used to fix a dye to the fabric. They also improve the take up quality of the fabric and help to improve colour and light fastness.

The discovery of chemical or synthetic dyes in the west in 19th century caused a massive blow in the Indian textile industry. There was a gradual decline in the use of natural dyes which were more expensive and more difficult to use in comparison to chemical dyes. Today there are only few areas where natural dyes are still used otherwise vegetable and mineral dyes are gradually being replaced by chemical ones. Some of the hand block printings are eco-friendly but some of them hazardous for nature. In recent years, many manufacturers have started using chemicals for increasing the fastness properties and durability of their dyed and printed articles but access use of chemicals is harmful for human beings as well as for nature.

Environmental factors play a vital role during selection of consumer goods including textiles all over the world. The German ban has severely affected the export of Indian textiles to these countries, as complying with the stringent eco-specifications has become an additional requirement apart from maintaining quality of the finished product. Most of the wet processing of textiles is done in decentralised sector; there is strong possibility of contamination of processed textiles with red-listed chemicals due to non-awareness among the small scale entrepreneurs about the newly introduced eco-standards and hazardous potential of various dyes and chemicals used in processing (The German Ban, 1996).

In recent years, due to increasing global warming the quality concept has assumed a key role in business. Consumer satisfaction has become of major importance for which purpose testing after every process is essential. Thus, in this new millennium, textile testing became more significant. It is well known that every product has an impact on the environment. However an average consumer does not know which product has less or more impact than the other one. Any product, which is made, used or disposed off in a way that significantly reduces the harm it would otherwise cause to the environment, could be considered as eco-friendly product. Slowly, consumers in India are taking lead in prompting manufacturers to adopt clean technologies to produce eco-friendly products.

Thus it becomes imperative for the manufacturers to gain insight regarding the permissible limits of particular chemical usage or its complete substitution. This would help them to contribute towards ecological growth and earn more profits in business.

Some of the chemical dyes were found to be associated with hazardous effect on human life, creating skin diseases and lungs problem. Environmentalists therefore, started searching for substitutes of chemical items which lead to the increase in use of natural dyes. The natural and vegetable dyes are more skin friendly. These colours have some depth which chemical dyes lack. These natural dyes are eco friendly and biodegradable if used without chemicals or within permissible limits of chemicals. While formulating eco-norms for the issuance of eco-labels, at present the use of five different classes of chemicals in textile production and processing are taken into consideration, i.e. released formaldehyde, pentachlorophenol (PCP), heavy metal traces, azo dyes (which release carcinogenic amines) and pH value. Block printing is the finest art of Rajasthan which has international demand since many years. Many exporters and manufacturing units however now are shifting towards the usc of automated machines to produce machine printed textiles. There is no doubt that factory or machine printed textiles are often cheaper and perhaps yield faster colours as compared to hand block printed fabrics. However, hand block printing represents a craft that provides a sustainable livelihood to rural artisans. It reflects human labour and the sensibilities of the craftsman, which no machine printed fabric, can ever do. In addition, it represents pride of our nation and therefore, the traditional craft must not be allowed to die.

Hence, present study has been designed to collect the information from owners and workers, related to different hand block printing techniques through survey. Their problems were also identified related to work space, working environment and working process and also assess whether seven hand block printing techniques used in Rajasthan (selected areas) are eco-friendly or not.

Therefore, the study has been focussed to check the quality of seven different hand block printed textile material by evaluating the colour fastness and to find out the chemicals and their severity, which are used for hand block printing in order to judge the eco-friendliness of these printed fabrics.

The eco-testing of block printed fabrics of Rajasthan has not been done before, so evaluation of these eco-parameters will help in creating awareness among owners and workers, as well as international markets may also be contacted for sale of eco-friendly hand printed items.

## APPLICATION IN SMART TEXTILES

Active wear needs to provide a thermal balance between the heat generated by the body while engaging in a sport and the heat released into the environment. Normal active-wear garments do not always fulfil this requirement. The heat generated by the body during strenuous activity is often not released into the environment in the necessary amount, thus resulting in a thermal stress situation. On the other hand, during periods of rest between activities less heat is generated by the human body. Considering the same heat release, hypothermia is likely to occur.

There are some commercial garments that possess microcapsule of PCM for example the registered mark OUTLAST ®, that help to prevent theses kinds of discomfort. In fact, in the case of heat generation PCM absorbs the energy thanks to the fusion of PCM and in the case of cold exposition release heat thanks to the solidification process. This system allows the thermoregulations of the garment and of it user.

## SHAPE MEMORY MATERIALS

### Principle

There are two types of Shape memory materials. The first classes are materials stable at two or more temperature states. In these different temperature states, they have the potential to assume different shapes, when their transformation temperatures have been reached. This technology has been pioneered by the UK Defence Clothing and Textiles Agency.

The other types of shape memory materials are the electroactive polymers which can change shape in response to electrical stimuli. In the last decade there have been significant developments in electroactive polymers (EAPs) to produce substantial change in size or shape and force generation for actuation mechanisms in a wide range of applications. In contrast to many conventional actuation systems, many types of EAPs are also capable of providing sensing functions. EAPs can provide a range of basic actuator mechanisms, force and displacement levels.

## Materials

Shape memory alloys, such as nickel-titanium, have been developed to provide increased protection against sources of heat. A shape memory alloy possesses different properties below and above the temperature at which it is activated. Below this temperature, the alloy is easily deformed. At the activation temperature, the alloy exerts a force to return to a previously adopted shape and becomes much stiffer. The temperature of activation can be chosen by altering the ratio of nickel to titanium in the alloy.

Cuprous-zinc alloys are capable of a two-way activation and therefore can produce the reversible variation needed for protection from changeable weather conditions. They will also react to temperature changes brought about by variations in physical activity levels.

Shape Memory Polymers have the same effect as the Ni Ti alloys but, being polymers, they will potentially be more compatible with textiles. The first SMPs were polynorborene-based with a Tg range of 35°C to 40°C developed by French CdF Chimie Company. Later, several classes of SMPs based on mix of styrene – butadiene - polyethylene Terephtalate - Polyetylene Oxyde – Polyurethane – Polycaprolactone –etc... were developed with tg from -46°C to 125°C for a widening of the types of application.

Electro active polymers EAPs are generally made up of high functionalised polymer. One of the most famous EAPs is the "gel robots" made up of poly 2 – acrylamido -2- methylpropane sulfonic acid that is fully researched for applications in the replacement of muscles and tendons.

## APPLICATIONS IN SMART TEXTILES

For clothing applications, the desirable temperatures for the shape memory effect to be triggered will be near body temperature.

In practice, a shape memory alloy is usually in the shape of a spring. The spring is flat below the activation temperature but becomes extended above it. By incorporating these alloys between the layers of a garment, the gap between the layers can be substantially increased above the activation temperature.

Consequently, considerably improved protection against external heat is provided. Polyurethane films have been made which can be incorporated between adjacent layers of clothing. When the temperature of the outer layer of clothing has fallen sufficiently, the polyurethane film responds so that the air gap between the layers of clothing becomes broader. This broadening is achieved if, on cooling, the film develops an out-of-plane deformation, which must be strong enough to resist the weight of the clothing and the forces induced by the movements of the wearer. The deformation must be capable of reversal if the outer layer of clothing subsequently becomes warmer.

Some active smart fibres contain electric conductive materials, Phase Change Materials PCM, and graphite particles, which can conduct electricity. In this way the resistance of the fibre is changeable along with the change of the fibre temperature due to change of fibre volume. As the material warms, it expands and reduces conductivity between graphic particles. These materials can automatically regulate the on/off of the electricity and keep the temperature stable. The shape memory alloys can also contribute to the miniaturization of equipment and systems, decrease the number of parts required and extend the life expectancy too due to the favourable fatigue properties of the alloy.

Considerable progress still needs to be made with EAP technologies before commercially viable applications. A multidisciplinary approach is essential for future developments. Applications such as fabrics and textile structures will require fibre-like EAP actuators and sensors in order to achieve effective integration. The large stimulated displacements that have been observed have encouraged new thinking in terms of both applications and designs. The natural ease of preparing and shaping such materials, coupled with their low mass and large displacements, opens up new approaches in many traditional areas as well as the potential to enable new technologies.

## CHROMIC MATERIALS

### Definition

Other types of intelligent textiles are those which change their colour reversibly according to external environmental conditions, for this reason they are also called chameleon fibres. Chromic materials are the general term referring to materials which radiate the colour, erase the colour or just change it because its induction caused by the external stimulus, as "chromic" is a suffix that means colour. Therefore we can classify chromic materials depending on the stimulus affecting them (in bold are indicated those used in textile)

Photochromic: external stimulus is light.

Thermochromic: external stimulus is heat.

Electrochromic: external stimulus is electricity.

Piezorochromic: external stimulus is pressure.

Solvatechromic: external stimulus is liquid or gas.

## MATERIALS AND APPLICATIONS IN SMART TEXTILES

Photocromic materials are generally reversible unstable organic molecules that change of molecular configuration with the influence of a special radiation. The molecular arrangement also perturbs the absorption spectra of the molecule and in consequences it colour. The applications in textile are intended to the fashion area and only a few for the solar protection. A T- Shirt made of photochromic prompted fabric was introduced to the market in 1989.

Thermochromic materials are those whose colour changes as a result of reaction to heat, especially through the application of thermochromic dyes whose colours change at particular temperatures. Two types of thermochromic systems that have been used successfully in textiles are: the liquid crystal type and the molecular rearrangement type. In both cases, the dyes are entrapped in microcapsules and applied to garment fabric like a pigment in a resin binder.

The most important types of liquid crystal for thermochromic systems are the so-called cholesteric types, where adjacent molecules are arranged so that they form helices. Thermochromism results from the selective reflection of light by the liquid crystal. The wavelength of the light reflected is governed by the refractive index of the liquid crystal and by the pitch of the helical arrangement of its molecules. Since the length of the pitch varies with temperature, the wavelength of the reflected light is also altered, and colour changes result. An alternative means of inducing thermochromism is by means of a rearrangement of the molecular structure of a dye, as a result of a change in temperature.

The most common types of dye which exhibit thermochromism through molecular rearrangement are the spirolactones, although other types have also been identified. A colourless dye precursor and a colour developer are both dissolved in an organic solvent. The solution is then microencapsulated and is solid at lower temperatures. Upon heating, the system becomes coloured or loses colour at the melting point of the mixture. The reverse change occurs at this temperature if the mixture is then cooled. However, although thermochromism through molecular rearrangement in dyes has aroused a degree of commercial interest, the overall mechanism underlying the changes in colour is far from clear-cut and is still very much open to speculation.

Toray Industries reported in 1987 the development of a temperature sensitive fabric by introducing microcapsules, diameter 3-4 mm to enclose heat sensitive dyes, which are resin coated homogeneously over fabric surface. The microcapsule was made of glass and contained the dyestuff, the chromophore agent (electron acceptor) and colour- neutralizer (alcohol etc.) which reacted and exhibited colour/ decolour according to the environmental temperature. SWAY was multicolour fabric, with basic 4 colours and combined 64 colours. SWAY can reversibly change colour at temperature greater than 5°C and is operable from - 40 to 80°C. The change of colour with temperature of these fabrics is designed to match the application, *e.g.* for ski-wear 11-19°C, women's clothing 13-22°C and temperature shades 24-32°C.

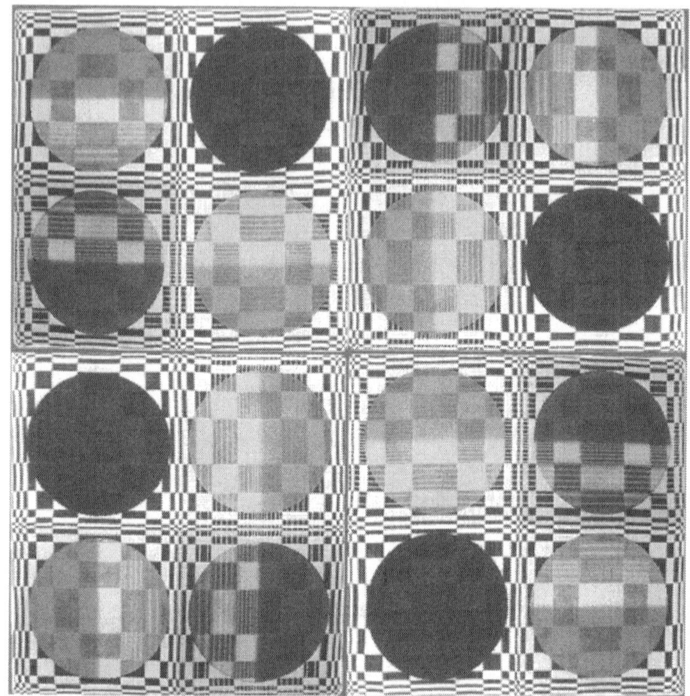

**Fig.** Uses of thermochromic inks by the International Fashion Machine

Other types of SFIT that use this effect are the electrically warming textiles (with Joules effect) which change colour with both the effect of warm and thermochromic materials. In addition to the changing of colour due to reaction to light or heat there are other chromic fibres presenting others characteristics. These fibres have raised the interest of people because of their surprising and interesting nature. Therefore, there is the problem that this "boom" will soon come to an end because these fibres are only considered to be a temporary fashion material. In order to establish these fibres in everyday life it is especially necessary to improve their endurance to light and to their accuracy.

Some of these fibres are those that present the phenomenon called solvate chromism, whose colour changes when in contact with a liquid, for example water. These materials are normally used for "design" swimsuits. Apart from this, the most important application for chromic materials is fashion, to create fantasy designs changing its colour depending on the volume of incident light.

## LUMINESCENT MATERIALS

The difference between chromic and luminescent materials is that the first one changes colour when the second one emits light thanks to a stimulus. There are several types of luminescent effects (in bold are indicated those used in textile):

- Photoluminescence: external stimulus is light. There are two types of photoluminescent materials the fluorescent and the phosphorescent. The only difference between the two is the time of emission.
- Opticoluminescence: conduction of light.
- Electroluminescence: external stimulus is electricity.
- Chemioluminescence: external stimulus is a chemical reaction.
- Triboluminescence: external stimulus is friction.

## TEXTILE DESIGN AND MANUFACTURING COMPONENT

The textile design and manufacturing component of the textile industry consists of specific companies who use dedicated CAD software or off-the-shelf CAD products for their product development processes. These companies represent textile design and textile manufacturing in South Africa.

### Specialized Computer-training Component

Tertiary education (also referred to as higher education) refers to all institutions that offer education that follows secondary education. These institutions include Universities, Technical colleges, Technikons and private providers of higher education. For the purpose of this study it was decided to concentrate on all Technikons in the tertiary education sector of South Africa offering the following courses:

- Textile Design.
- Fashion Design and Technology
- Clothing Production.

### Apparel Design and Manufacturing

The population of the apparel design and manufacturing component consisted of all manufacturers currently using CAD/CAM systems supplied by three specific CAD/CAM technology suppliers in South Africa. The three CAD systems (in alphabetical order) are:

- Gerber, supplied by Intamarket.
- Lectra, supplied by Lectra systems of SA.
- Polygon, supplied by Midcomp.

Each of the above-mentioned suppliers of CAD technology, supply CAD and CAM software packages relative to product development, product presentation and manufacture. The above-mentioned suppliers sell CAD/CAM technology to apparel manufacturers in the selected geographical regions of South Africa. The distribution of apparel designers and manufacturers using CAD/CAM systems from the three identified suppliers in South Africa.

The three CAD/CAM suppliers were approached and asked for permission to access their client list for the three identified prominent geographical manufacturing areas in South Africa. These apparel designers and manufacturers are currently using CAD/CAM equipment for product development and production. The population of the apparel design and manufacturing component and the method by which the sample was selected.

A *quota sampling method* was used to select the sample. Quota sampling is a simplified form of stratified random sampling. The method amounts to the formation of cells (or homogenous sub-populations) by using control characteristics for which the figures of the population are available. The sample sizes can be taken from each cell, and sample elements are then collected at random.

## APPAREL DESIGN AND TECHNOLOGY COMPONENT

The apparel design and technology component consisted of nine organizations selected from each software supplier, three from each specified geographical region. A table of random numbers was used to select the manufacturers.

A total of 27 manufacturers were selected. Only three Polygon users are situated in the Cape region. One of the selected candidates could not be reached, and subsequently only two manufacturers where approached. Of the initial 27 selected manufacturers, only 26 were contacted. 5.4.2 Textile design and manufacturing

The population of the textile design and manufacturing component consisted of manufacturers who have an established relationship with the Department of Textile Design of Technikon Pretoria. They allow the students of the department into their production environment as visitors, or as new recruits.

These manufacturers are familiar with the level of expertise of the student who finishes his/her course in the Department of Textile Design at Technikon Pretoria. These manufacturers produce a variety of textile products.

The types of CAD and CAM systems used by the textile segment of the textile industry vary greatly. A number of generic graphic design computer packages are also used. No specific CAD systems, dedicated to the textile segment of the textile industry, or suppliers of CAD and CAM systems were identified, as opposed to the case with the apparel design and manufacturing component. The computer technology that the textile design and manufacturing population use is a good representation of the type of CAD equipment currently in use by the textile industry at large. It is also the computer-related design and production environment the students of the Department of Textile Design of Technikon Pretoria could find themselves in, on completion of their studies. These organizations represent textile design, surface design and textile manufacturing; and include large, medium and small manufacturers.

A *purposive sampling method* was used for selecting the sample for the textile design and manufacturing component. This method allows the researcher to identify the selection based on the knowledge the researcher has of the population. The segment of the population was chosen based on the relationship these design studios and manufacturers have with the Department of Textile Design at Technikon Pretoria. The sample therefore consisted of fourteen specific textile design studios and textile manufacturers. The number of manufacturers selected and the region in which these design studios and manufacturers are situated.

**Specialized Computer-training**

Technikons in South Africa that offer courses in fashion and textile design were considered to represent tertiary education. The segment of the population selected for the sample consisted of Technikons where the specific fashion and textile design courses are presented in the departments of Fashion and Textile Design.

These courses are:

- Fashion Design and Technology;
- Production Technology; and
- Textile Design and Technology.

The Departments of Fashion- and Textile Design were the population the population for the training component for the empirical study. The population of the specialized computer-training component, and the sampling method.

Technikons. As in the textile design and manufacturing component, a *purposive sampling method* was used. The selected Technikons (in alphabetical order) are:

- Border Technikon
- Cape Technikon
- Durban Technikon
- Eastern Cape Technikon
- ML Sultan
- Peninsula Technikon
- PE Technikon
- Technikon North-West
- Technikon Witwatersrand
- Vaal Triangle Technikon

**Research Method**

The survey was conducted by directly approaching the companies or departments in the various research samples. A specified contact person was established in all companies and/or departments. The interview schedule was forwarded directly to the identified contact person, via Internet or facsimile. This method of survey was possible due to the low number of interviewees. A 100% response rate was achieved. All interview schedules were sent out in January and February 2001. Once the correct contact person had been determined the person was approached telephonically. Background information and an explanation on the nature of the research were disclosed.

A few interview schedules were not returned in the specified time. These companies/departments were contacted telephonically. When no response was received after requests to return the interview schedule, the interviewee was contacted and together with the interviewer, the interview schedule was completed telephonically.

# 8

# Textile and Apparel Manufacturing

Industries in the textile mills sector include establishments that transform a basic, natural or synthetic fibre into a product like yarn or fabric that is further manufactured into usable items like apparel, sheets, towels, and textile bags for individual or industrial consumption.

Further manufacturing may be performed in the same establishment and classified in this subsector, or it may be performed at a separate establishment and be classified elsewhere in manufacturing.

The main processes in this sector include preparation and spinning of fibre, knitting or weaving of fabric, and the finishing of the textile. Major industries such as preparation of fibres, weaving of fabric, knitting of fabric, and fibre and fabric finishing are uniquely identified in this subsector. Texturizing, throwing, twisting, and winding of yarn contains aspects of both fibre preparation and fibre finishing and are classified with preparation of fibres rather than with finishing of fibre.

Industries in the apparel manufacturing sector include establishments with two distinct manufacturing processes: cut and sew (*i.e.*, purchasing fabric and cutting and sewing to make a garment) and the manufacture of garments in establishments that first knit fabric and then cut and sew the fabric into a garment.

The apparel manufacturing sector includes a diverse range of establishments manufacturing full lines of ready-to-wear apparel and custom apparel. These include apparel contractors that perform cutting or sewing operations on materials owned by others, jobbers performing entrepreneurial functions involved in apparel manufacture like knitting, and tailors that manufacture custom garments for individual clients.

Establishments in the leather sector transform hides into leather by tanning or curing and fabricating the leather into products for final consumption. Also included is the manufacture of similar products from other materials, including products (excluding apparel) made from leather substitutes like rubber, plastics, or textiles. Rubber footwear, textile luggage, and plastics purses or wallets are examples of leather substitute products included in this group.

## LEATHER APPAREL MANUFACTURING

This industry comprises establishments primarily engaged in manufacturing cut and sew gloves (excluding rubber, metal, and athletic gloves) and mittens from purchased fabric, fur, leather, or from combinations of fabric, fur, or leather.

Jobbers, who perform entrepreneurial functions involved in glove and mitten manufacture, including buying raw materials, designing and preparing samples, arranging for gloves and mittens to be made from their materials, and marketing finished gloves and mittens, are included.

## HAT, CAP, AND MILLINERY MANUFACTURING

This industry comprises establishments primarily engaged in manufacturing cut and sew hats, caps, millinery, and hat bodies from purchased fabric.

Jobbers, who perform entrepreneurial functions involved in hat, cap, and millinery manufacture, including buying raw materials, designing and preparing samples, arranging for hats, caps, and millinery to be made from their materials, and marketing finished hats, caps, and millinery, are include

## MEN'S AND BOYS' CUT AND SEW SHIRT (EXCEPT WORK SHIRT) MANUFACTURING

This industry comprises establishments primarily engaged in manufacturing men's and boys' outerwear shirts from purchased fabric. Men's and boys' shirt (excluding work shirt) jobbers, who perform entrepreneurial functions involved in apparel manufacture, including buying raw materials, designing and preparing samples, arranging for apparel to be made from their materials, and marketing finished apparel, are included.

Unisex outerwear shirts, such as T-shirts and sweatshirts that are sized without specific reference to gender (*i.e.*, adult S, M, L, and XL) are included in this industry.

## TEXTILE MANAGEMENT

Textile management is a professional industry concerned with the make and management of clothing and apparel. The textile industry is often combined with fashion merchandising and design, and universities that prepare students to enter a career in textile management usually have one department for textile, fashion and design.

The goal of these programs is to create professionals who will enter the field able and willing to create and interpret knowledge of clothing materials to better serve the industry. A degree in textile management may be granted as a business administration degree, as many textile students hope to become textile manufacturers or other business owners who specialize in textile production. For these careers, some knowledge of business is required, so students studying textile management will usually take at least a few business courses. Textile marketing is another popular field in the textile industry, and requirements include studying the clothing market, participating in hands-on research projects, developing products and coordinating sales efforts with clothing stores. Students interested in this aspect of the textile industry will focus heavily on marketing courses, and they will learn quickly how to apply principles of public relations and sales to the clothing industry. Another area of textile management is the production of textile fabric. Students interested in this field should be hands-on learners interested in creating tangible products. Textile technology is constantly changing, so students interested in this area should be willing to be on the lookout for new trends in the clothing industry. The creation of textile fabric includes producing yarn and any of several kinds of fibres. People who are drawn to knitting and crocheting might find an interest and talent in this aspect of textile management.

Other areas of the textile industry where professionals who receive education and training in this discipline may work include merchandising, inventory control, sales promotion, public relations and human resources. Many career opportunities are present for students interested in working with textiles, and because of the broad nature of the field, there is opportunity for movement within the industry during a lifelong career. It is a fast-paced, diverse and exciting career field, although it can also be competitive, depending on which side of the industry you are interested in.

Opportunities for employment in textile management can be pursued with retailers, manufacturers, market research firms and product information venues. Typical tasks of a textile professional may include studying fashion changes and the corresponding impact on the market, developing new apparel lines, studying customer behaviour and purchase cycles, analysing statistics, and operating and organizing retail stores.

## DIFFERENT TYPES OF TEXTILE COMPANIES

Textiles are usually defined as any type of item that is made using yarn or thread to createsome kind of cloth or fabric. There are many different kinds of textile companies that make up the worldwide textile industry, many of them specializing in one or two specific textile product lines. In some cases, the focus of the company is on a residential clientele, while others prefer to target commercial customers as their core client base. Textile companies often base their reputation and product line of the creation and sale of what was once known in the United States as a home furnishing line. Essentially, this approach to textile fabric manufacture would be aimed at producing items like draperies, bedding, tablecloths, and other fabric products that were common to the task of decorating the home. Most experts only include textiles that are stand-alone products in this category. Fashion textile companies tend to devote themselves to the design, manufacture, and marketing of clothing apparel. This can include the production of solid lots of material like denim or cotton sheeting that is sold to a customer who uses the fabric to create specific items of clothing. Synthetic blends may also be created and sold that are ideal for use in producing winter coats, suits, dresses, swimsuits, or any other type of apparel worn by men, women, or children.

Textile companies that focus on the creation of upholstery materials are often identified as industrial textile producers. This classification can include the production of faux leather using a mix of synthetic and natural fibres, heavy fabrics for use in furniture upholstery, and even refining the yarn that is eventually used in the creation of various types of carpeting. While the inclusion of the manufacturing of materials for carpets was once considered to be separate from the textile industry, it is not unusual for carpeting manufacturers to be including in the family of textile manufacturers today.

Just as there are different forms of textile production, the different textile companies in operation today are likely to employ a wide range of quality standards to their finished goods. In general, textile management teams will seek to comply with any minimum standards set in place where the mills and plants are established, as well as meet the standards necessary to allow sale of the goods in various countries. While the United States, Great Britain, and Canada were once leaders in the textile industry, the vast majority of textile products are now produced outside those three nations.

## DIFFERENT TYPES OF TEXTILE MACHINE

*Textile Machine Manufacturers on wiseGEEK*:

- Knitting is traditionally a textile production that is completed by hand with a needle or a crochet hook, but industries also incorporate large knitting machines as well. Crocheting is another type of textile production that would fall under this category.
- In some cases, fabric may be cut and sown by hand with a needle and thread. Other times, textiles could be produced on a sewing machine. Large manufacturing facilities often generate cloth goods on computerized equipment specifically designed for making a particular item.

*Textile Weaving Machine on wiseGEEK*:

- Silk fabric is woven through this method, as are many commercial textiles and linens. There are hand-operated shuttle weaving machines as well, but most non-electric weaving machines tend to be used by trained professionals for artisanal weaving.
- One of the earliest adopters of both directed and computerized machines was the textile industry. They created weaving machines

that first used an unskilled operator and then a computer, instead of a skilled weaver.

## TEXTILE PRINTING MACHINE

*Textile Printing Machine on wiseGEEK*:

- For drying, there are steam dryers, drum dryers, conveyor dryers, tumble dryers. For printing, the types of textile machine that are used include a rotary screen printing machine, a heat transfer printer, or a transfer printing calender.
- Poorly maintained factories or equipment can be especially dangerous for workers, and textile mills can also generate environmental pollution in the form of emissions from their power plants or the release of chemicals used in textile manufacturing. Staff at a textile mill includes production workers who run the machines, maintenance crews who keep the machines and the factory in good order, and supervisors who have scheduling and determine what will be produced when.

## FUNCTION WISE CLASSIFICATION OF TEXTILE PROCESSING MACHINERY

Right from the time the textile item such as yarn or cloth is received from the manufacturing section, *i.e.*, the spinning winding or weaving department, it is to be subjected to treatment by a series of machineries, according to the specific plan. Each of the machine conducts the textile one step forward towards the end product, which may be the ready-to-use yarn or the fabric. Although it would really suffice to review the latest model of each of the processing machines in order to have a proper understanding of the functions, it is deemed necessary to review the items related to various processing functions. The following gives the functional classification of the various items.

- Pre-preparatory Equipments:
  - Singeing machines - singe preparatory sewing machines, batching machines, A frames.
  - Desizing machines - What is desizing any way?
  - Mercerising Machines - woven fabric - chain mercerising machines and chainless mercerising machines Knit fabric - Tubular Mercerising machines, Yarn Mercerising machines
  - Scouring and Bleaching machines: Open width Scouring machine
- Dyeing Machinery
- Modern colour kitchen with automated chemical dispensing systems
  - Jiggers - jumbo jiggers, hooded jiggers, JT-10 Jiggers - woven fabric open width dyeing machines.
  - Cabinet dyeing machines - hank and loose fibre dyeing machine.

  - HT/HP Cheese and Cone dyeing machines - vertical type - horizontal type
  - Jet Dyeing machines - HT/HP - woven and knit goods synthetic and its blends dyeing machine.
  - Soft Flow Dyeing machines - Hosiery material dyeing machines - atmospheric and HT/HP types.
  - Semi continuous Dyeing Machines - woven and knit goods
  - Continuous Dyeing Machines - woven and knit goods
  - Garment Dyeing Machines
- Print-preparatory Machineries
- Printing Machinery
- Post Printing machinery
- Drying and finishing machineries
- Machines for Physical treatment: Decatising machine, Continuous Pressing machines, Crabbing Machines, Shearing machines, Sueding machines, Shrinking machines
- Machines for fabric make up and packing: Make up machines
- Yarn Processing machineries: Hank Yarn Dyeing - Cabinets
- Machinery for Special Effects
- Sample Dyeing Machines
- Machinery not falling under the above categories
- Quality Control - Testing Machines.
  - Physical Testing machines
  - Chemical Testing machines

These machines are being utilized to achieve the objectives as described in the respective terminology used for the particular class. Some of the machinery items are being used for functions covered by more than one classes, owing to the over lapping nature of the functions.

## SCOPE OF TEXTILE TRADE

Textile design is an artistic craftsmanship that associates itself in furnishing creative, stylish and contemporary designs.

Textile design requires special skills and imaginations to create innovative designs.

*The skills required for effective textile design making are*:

- Creative imagination
- Technical skills
- Commercial awareness of textile industry with continuous updation of fashion trends
- Research and data handling
- Critical evaluation and interpretation of materials
- On the spot decision making ability

## KEY ASPECTS OF TEXTILE DESIGN

The core area of textile designing involves the following:

- Designing fabric by using different technique comprising of printing, weaving, ornamenting fabric, print technique, tracing embroidery and colour detailing.
- Providing support to the clients to visualize the design.
- Helping the clients to correct samples while executing prototypes.

Main activities in textile design

Textile designing involves the following basic activities.

- Creating sets of design samples
- Coordinating with clients to plan and develop designs
- Experimenting with colour, fabric and texture.
- Conceptualizing new innovative designs
- Attending trade shows regularly
- Keeping up with emerging fashion trends

## DESIGN STUDIOS

A textile design studio comprises of a group of textile designers and is equipped with computers and design software to provide quality service and maintain professional standards.

*Range of studio design products*: The different range of products consists of bed linens, table linens, floor covering, cushions and curtains and many more.

*Designing Activities*:

- Textile designing
- Designing stalls for exhibitions
- Designing showrooms

A design studio after creating its innovative product design, ultimately puts them up for sale to its world wide buyers. Career in textile designing with growing market potential for textile products there is a wide range of career options in this field. Textile design comprises both surface design and structural design. Textile designers usually handle embroidery designs, print, weave and texture. Textile designers need to have detail information about textile fibres and different methods of textile design..

## TEXTILE AND APPAREL INDUSTRY "AN ANALYSIS"

## STRENGTHS

- Indian Textile Industry is an Independent and Self-Reliant industry.
- Abundant Raw Material availability that helps industry to control costs and reduces the lead-time across the operation.
- Availability of Low Cost and Skilled Manpower provides competitive advantage to industry.

- Availability of large varieties of cotton fibre and has a fast growing synthetic fibre industry.
- India has great advantage in Spinning Sector and has a presence in all process of operation and value chain.
- India is one of the largest exporters of Yarn in international market and contributes around 25% share of the global trade in Cotton Yarn.
- The Apparel Industry is one of largest foreign revenue contributor and holds 12% of the country's total export.
- Industry has large and diversified segments that provide wide variety of products.
- Growing Economy and Potential Domestic and International Market.
- Industry has Manufacturing Flexibility that helps to increase the productivity.

## TEXTILE INDUSTRY

- Textile Industry concerns
- Textile Sectors
- Handicrafts Sector
- Handloom Sector
- Textile Industry India
- Textile Investment Policies

## PAINTINGS

- Ragamala Painting
- Landscape Paintings
- Renaissance Paintings
- Da Vinci Painting
- Monalisa
- Faux Painting

## STONE WORK

- India Marble Stone
- Stone Handicrafts
- Marble Write up
- Stone Carving
- Stone Write up
- Marble Sculptures

## CRAFT WORLD

- Crafts
- Art and Crafts
- Valentine Crafts
- Kids Crafts

- Crafts Show
- Indoor Craft
- Halloween Craft

**WEAKNESSES**

- Indian Textile Industry is highly Fragmented Industry.
- Industry is highly dependent on Cotton.
- Lower Productivity in various segments.
- There is Declining in Mill Segment.
- Lack of Technological Development that affect the productivity and other activities in whole value chain.
- Infrastructural Bottlenecks and Efficiency such as, Transaction Time at Ports and transportation Time.
- Unfavourable labour Laws.
- Lack of Trade Membership, which restrict to tap other potential market.
- Lacking to generate Economies of Scale.
- Higher Indirect Taxes, Power and Interest Rates.

**OPPORTUNITIES**

- Growth rate of Domestic Textile Industry is 6-8% per annum.
- Large, Potential Domestic and International Market.
- Product development and Diversification to cater global needs.
- Elimination of Quota Restriction leads to greater Market Development.
- Market is gradually shifting towards Branded Readymade Garment.
- Increased Disposable Income and Purchasing Power of Indian Customer opens New Market Development.
- Emerging Retail Industry and Malls provide huge opportunities for the Apparel, Handicraft and other segments of the industry.
- Greater Investment and FDI opportunities are available.

**THREATS**

- Competition from other developing countries, especially China.
- Continuous Quality Improvement is need of the hour as there are different demand patterns all over the world.
- Elimination of Quota system will lead to fluctuations in Export Demand.
- Threat for Traditional Market for Powerloom and Handloom Products and forcing them for product diversification.
- eographical Disadvantages.
- International labour and Environmental Laws.
- To balance the demand and supply.
- To make balance between price and quality.

## TEXTILE INDUSTRY INITIATIVES

The primary objective is to create an environment that helps the industry to compete on the global basis.

- To build the environment that will focus on: Wealth Creation, Infrastructure Development, Training, Technological Development, and Poverty Alleviation etc. to enhance the sector performance.
- Transportation infrastructure should be improved to access the untapped market that would be beneficial to reduce the transportation cost
- There should be optimized distribution network and supply chain management.
- Developing technologies to recycle natural resources to produce new products and ensure waste minimization, product durability and reliability.
- Encourage Private Sector for Partnership and collaboration.
- More Training centers should be opened to train the workforce and awareness of new technology and trends should be increased among manpower.
- There should be more Special Economic Zones should be opened with focus on development of Textile Industry.
- Labour Laws should be more liberalized and favourable that will help to make them more productive.

## QUALITY CONTROL OF APPAREL AND TEXTILE

For every industry or business, to get increased sales and better name amongst consumers and fellow companies it is important to maintain a level of quality. Especially for the businesses engaged in export business has to sustain a high level of quality to ensure better business globally. Generally quality control standards for export are set strictly, as this business is also holds the prestige of the country, whose company is doing the export. Export houses earn foreign exchange for the country, so it becomes mandatory to have good quality control of their products.

In the garment industry quality control is practiced right from the initial stage of sourcing raw materials to the stage of final finished garment. For textile and apparel industry product quality is calculated in terms of quality and standard of fibres, yarns, fabric construction, colour fastness, surface designs and the final finished garment products. However quality expectations for export are related to the type of customer segments and the retail outlets.

There are a number of factors on which quality fitness of garment industry is based such as - performance, reliability, durability, visual and perceived quality of the garment. Quality needs to be defined in terms of a particular framework of cost. The national regulatory quality certification and international quality programmes like ISO 9000 series lay down the broad quality parameters based on which companies maintain the export quality in the garment and apparel industry.

*Here some of main fabric properties that are taken into consideration for garment manufacturing for export basis*:

- Overall look of the garment.
- Right formation of the garment.
- Feel and fall of the garment.
- Physical properties.
- Colour fastness of the garment.
- Finishing properties
- Presentation of the final produced garment.

## SOURCING OF FABRICS

There are certain problems that could be faced by garment manufacturers when sourcing for certain fabrics, so precautions should be taken for it beforehand to minimize the problems. The garment exporters source cotton fabrics mainly from handloom sectors, powerlooms and mills. Each of these sectors presents their own unique set of problems to the garment exporters. Sourcing cotton from handloom sectors might present some set of problems like colour variation, missing ends and picks, irregular weaves and unreliable supplies. However, the handloom sector is significant source of heavier cotton. Common problems faced in powerloom cotton sourcing are broken ends and reed marks, thick and thin places, difference in width and massive variation in costing. The major problem in mill-made fabric sourcing is to meet huge demands from the mills. Fabrics have to be ordered well in advance in mills and the long time taken for producing the fabric is a matter of concern for garment exporters. Mills generally hesitate to take small orders which pose a problem for small scale exporters.

- It is not that sourcing problems which only confined to cotton fabrics, but also to other fabrics as well. In silk garment industry there are some sorts of problems faced by silk garment exporters. Some of the problems that could be faced by silk garment exporters are as follows:
- Shortage of imported silk yarns in the quantities required, as a result delivery is delayed.
- Silk material is very vulnerable to stains during manufacturing process as well as stocking, staining results in rejection so a lot of care has to taken during these procedures.
- Roll length of the silk yarn is often insufficient.
- Colour fastness of dyed silk material is sometimes not satisfactory.
- There are also chances of warp breakage.
- Basic Thumb Rules for Garment Exporters

- For a garment exporter there are many strategies and rules that are required to be followed to achieve good business. The fabric quality, product quality, delivery, price, packaging and presentation are some of the many aspects that need to be taken care of in garment export business. Quality has to be taken care by the exporter, excuses are not entertained in international market for negligence for low quality garments, new or existing exporters for both it is mandatory to use design, technology and quality as major upgradation tools.
- Apart from superior quality of the garment, its pricing, packaging, delivery, etc has to be also taken care of.
- The garment shown in the catalogue should match with the final garment delivered.
- It is important to perform according to the promises given to the buyer, or else it creates very bad impression and results in loss of business and reputation.
- In international market, quality reassurance is required at every point.
- Proper documentation and high standard labels on the garment are also important aspects as these things also create good impression.
- Timely delivery of garments is as important as its quality.
- If your competitor has the better quality of garment in same pricing, it is better to also enhance your garment quality.
- Before entering into international market, garment exporters have to carefully frame out the quality standards, or else if anything goes wrong it could harm the organization. And after that strictly follow it.
- The garment quality should match the samples shown during taking the orders.
- The garment exporters should know to negotiate a premium price after quality assurance is done.
- Quality is a multi-dimensional aspect. There are many aspects of quality based on which the garment exporters are supposed to work.
- Quality of the production.
- Quality of the design of the garment.
- Purchasing functions' quality should also be maintained.
- Quality of final inspection should be superior.
- Quality of the sales has to be also maintained.
- Quality of marketing of the final product is also important as the quality of the garment itself.

*There are certain quality related problems in garment manufacturing that should not be overlooked*:

- *Sewing defects*: Like open seams, wrong stitching techniques used, same colour garment, but usage of different colour threads on the garment, miss out of stitches in between, creasing of the garment, erroneous thread tension and raw edges are some sewing defects that could occur so should be taken care of.

- *Colour effects*: Colour defects that could occur are - difference of the colour of final produced garment to the sample shown, accessories used are of wrong colour combination and mismatching of dye amongst the pieces.
- *Sizing defects*: Wrong gradation of sizes, difference in measurement of a garment part from other, for example- sleeves of 'XL' size but body of 'L' size. Such defects do not occur has to be seen too.
- *Garment defects*: During manufacturing process defects could occur like - faulty zippers, irregular hemming, loose buttons, raw edges, improper button holes, uneven parts, inappropriate trimming, and difference in fabric colours.

## CONCLUSION

Quality is ultimately a question of customer satisfaction. Good Quality increases the value of a product or service, establishes brand name, and builds up good reputation for the garment exporter, which in turn results into consumer satisfaction, high sales and foreign exchange for the country. The perceived quality of a garment is the result of a number of aspects, which together help achieve the desired level of satisfaction for the customer. Therefore quality control in terms of garment, pre-sales service, posts -sales service, delivery, pricing, etc are essentials for any garment exporter.

One of the keys to producing good quality merchandise is an in-process quality control programme. Although it is possible to control your outgoing quality with only a good final audit, it is NOT recommended to simply rely on that approach. Unless you install an effective in-process quality control programme, your cost of excessive seconds and repairs may be high. It is far more effective to correct the problem at the operator level, then after the garment is completely assembled, pressed, packages and prepared for shipping. Being able to deliver your merchandise on time is important to your customers. Good in-process controls help assure that the final audit runs smoothly and allows for timely delivery. You certainly do not want to learn in your final audit about problems that could have easily been fixed if detected earlier.

The primary purpose of the in-process auditing is to identify problems as early as possible. A problem may be caused by the operator, the machine, or other factors. The inline audits will help you find specific problems in production. The only way to fix a problem is to find the problem. It is important to find errors as quickly as possible so that they can be corrected as fast as possible. Some companies will do their own inline inspections and others will utilize inspection services.

## QUALITY ASSURANCE ELEMENTS

In QA elements we should consider the quality of all raw aterials, right from the processing of fabric or yarn to all the dyes and chemicals used and in QC elements we should test the quality of finished products for the required level of standards. In both cases the following Textile Testing Methods can be applied for.

## RAW MATERIALS AND OTHER INPUTS:

- Water Quality
  - Hardness
  - bi-carbonate
  - Carbonates. sulphates and chloride
  - TDS (Total Dissolved Solids)
  - Turbidity
  - Other metals -iron, copper

## QUALITY OF AUXILIARY CHEMICALS USED.

- Quality of surfactants like wetting, scouring agents
  - Wetting time of wetting agents
  - Solid content
  - Ionic nature
  - Cloud point of non-ionic products
  - Foaming characters
- Quality of sequestering agents
  - Chelating power
  - pH
  - Solid content
- Quality of leveling/dispersing agents
  - Ionic nature
  - Concentration required
  - Solid content
  - Suitability for metal complex dyes
- Quality of Peroxide Stabilizers
  - Solid content
  - Concentration required for bleaching
- Quality of peroxide killers
  - Whether it is an enzyme or inorganic product
  - Concentration required for treatment
  - If it is enzyme, get the test method from the manufacturer himself.
- Quality of Cationic dye fixing agents
  - Solid content

  - Free Formaldehyde content
  - Degree of Change in tone of a shade
  - How much light fastness is affected?
- Quality of Softeners
  - Ionic nature
  - Degree of softness acquired for a standard concentration
  - Degree of tone change of a shade
  - Wash Durability. Solid content
- Determination of Acid Value of a product?
  - Detailed method to find out the acid value of a product
- Determination of Polymer content of a binder?
  - Detailed method to find out the polymer content of a binder.
- Quality of Dyestuff:
  - Dyestuff properties and selection method for a combination

## QUALITY OF BASIC CHEMICALS USED AND ITS CONCENTRATION

- Caustic soda - purity and concentration percentage
- Soda Ash - purity and concentration
- Acetic Acid, Hydrochloric acid, Sulphuric acid, Oxalic acid - purity and concentration
- Hydrogen Peroxide - purity and concentration
- Common Salt - purity, hardness of a particular concentration
- Sodium Hypochlorite - concentration
- Test for Iron and Copper in cotton fibre

## QUALITY REQUIREMENTS OF GRIEGE (UN-PROCESSED) TEXTILE GOODS

- Greige Cotton fabric/Yarn quality requirements
- Cotton Maturity Tests
- Greige Wool fabric/Yarn quality requirements

## QUALITY CONTROL ELEMENTS

- Physical Properties:
  - Appearance - colour and uniformity of finished yarn or fabric - visual assessment only.
  - Tensile strength of finished yarn
  - CSP of finished yarn
  - Tear strength of finished fabric
  - GSM of fabric of finished
  - Shrinkage Test of Woven fabric and Knit fabric.
- Finished goods' Qualities:

– What is a Grey Scale and Staining Scale? How Fastness grading is done?
– Fastness to Washing
– Fastness to Crocking or Rubbing fastness
– Fastness to water
– Fastness to perspiration
– Fastness to peroxide bleach
– Fastness to Chlorine or Chlorinated Pool water fastness
– Fastness to Mercerising
– Fastness to Hydrolysis
– Fastness to Dry-cleaning
– Fastness to light
– Pilling Tests
– What is whiteness Index
– What is total colour difference or DE
– Method of Detecting Starch and PVA
– Method of testing desizing efficiency
– Measurement of Absorbency
– Antimicrobial Tests
– Flammability Test
– Detection and estimation of Formaldehyde content in Textiles

Like biological evolution, textiles also have gone through a lot of metamorphosis to reach the present day level. It is surprising to note that from time immemorial biological processing of textiles have taken place in one way or the other to bleach, colour and print. In this article we are focusing on various enzymes, eco-friendly processes and eco balancing.

## ROLE OF BIOTECHNOLOGY IN TEXTILE PROCESSING

The major areas of applications of biotechnology in textile industry are,

- Improvement of plant varieties used in production of textile fibres and in fibres and in fibre properties
- Improvement of fibres derived from animals and health care of animals
- Novel fibres from biopolymers and genetically modified microorganisms
- Replacement of harsh and energy demanding chemical treatments by Environment friendly routes to textile auxiliaries such as dyestuffs
- Novel uses for enzymes in textile finishing
- Development of low energy enzyme based detergents
- New diagnostic tools for detection for adulteration and quality control of textiles
- Waste managements

## ROLE OF ENZYMES IN TEXTILE PROCESSING

Enzymes are large protein molecules made up of long chain amino acids which are produced by living cells in plants, animals and microorganism such as bacteria of fungi.

Enzymes are secretions of living organisms, which catalyze biochemical reactions. Enzymes are biocatalysts without which no life in plant or animal kingdom can be sustained.

# Bibliography


A.K. Dubey and Anil Kumar.: *Engineering Mechanics1*, New Age Publications, Delhi, 2008.

Alimuddin Khan.: *Basics of Classical Mechanics*, Anmol Publications, Delhi, 2008.

Arvind M. Nawale: *Anita Desai's Fiction: Themes and Techniques*, B.R. Publication, Delhi, 2011.

Avtar Singh Bimbraw: *Agro-techniques for Umbelliferous Medicinal and Aromatic Plants of India*, International Book Distributing Co Publication, Delhi, 2006.

Barnes W. McCormick.: *Aerodynamics, Aeronautics, and Flight Mechanics*, Wiley India, Chennai, 2010.

C. Emmanuel, Fr. S. Ignacimuthu,s.j. and S. Vincent: *Applied Genetics: Recent Trends and Techniques*, MJP Publication, Delhi.

Chinmoy Goswami and Abhijit Paintal: *A Textbook of Laboratory Techniques*, Wisdom Press Publication, Delhi, 2011.

Eschenauer.: *Applied Structural Mechanics*, Springer Publications, Chennai, 2010.

Fomin, V.M., Kiselev, S.P., Vorozhtsov.: *Fluid Mechanics*, Jaico Publications, Delhi, 2009.

G D Arora.: *Analytical Mechanics*, Sarup Publications, Delhi, 2007.

Goutom Sharma and Julie Barooah: *Textile Tradition of Assam : Collection of Purbajyoti Sangrahalaya*, Indian Museum, 2006.

Gregory.: *Classical Mechanics*, Cambridge University Press, Delhi, 2011.

James Turley: *Advanced 80386 Programming Techniques*, Tata McGraw-Hill Publication, Delhi, 2008.

Jatinder S. Bedi: *Economic Reforms and Textile Industry: Performance and Prospects* , Commonwealth Publication, Delhi, 2002.

Khaja Mohtesha Muddin: *A Practical Manual of Veterinary Andrology and Reproductive Techniques*, Satish Serial Publicationg House, 2015.

M. Soundarapandian: *Textile Industry Under Globalization*, Dominant Publication, Delhi, 2004.

Mahinder Singh: *A Textbook of Analytical Chemistry: Instrumental Techniques*, Dominant Publication, Delhi, 2005.

Michael Armstrong: *A Handbook of Management Techniques: A Comprehensive guide to Achieving Managerial Excellence and Improved Decision Making*, Viva Books Publication, Delhi, 2010.

Nikulsinh M. Chauhan: *Agricultural Extension: Modern Tools and Techniques,* Biotech Books Publication, Delhi, 2016.

O.P. Sharma, Narendra Singh and Ruchira Shukla: *Advanced Management Techniques*, Agrotech Publication, Delhi, 2012.

P. Venkatesan: *A Textbook of Biomedical Laboratory Techniques*, Atlantic Publication, 2011.

Rajaram Jaipuria: *Textile Legend Unravels*, Jaipuria Publication House, 2007.

S. Baluchamy: *Women in Textile Industry: Problems and Prospects*, Mittal Publication, Delhi, 2012.

S.P. Tiwari, S. Rajagopal and Usha Rani Mehra: *Analytical Techniques in Animal Nutrition*, Satish Serial Publication, Delhi, 2012.

Sangeeta Gupta: *Accounting and Statistical Techniques*, Pointer Publication, Delhi, 2009.

Shashi Kant Yadav.: *Textbook of Classical Mechanics*, Discovery Publications, Delhi, 2011.

Shuji Uchikawa: *Indian Textile Industry: State Policy, Liberalization and Growth*, Manohar Publication, Delhi, 2004.

Subha Gnaguly: *A Laboratory Manual on Virology and Tissue Culture Techniques*, Narendra Publication House, Delhi, 2014.

Sukdeb Nandi: *Advanced Molecular Diagnostic Techniques: Principles and Applications,* Satish Serial Publication House, Delhi, 2016.

Sumit Narula and R.K. Jain: *An Introduction to Journalism: Principles and Techniques*: , Regal Publication, Delhi, 2012.

Tanya Jain: *Textile Designing : Theory and Concept*, Yking Books, Delhi, 2012.

V.K. Agarwal and S.D. Upadhyaya: *Agrotechniques of Medicinal and Aromatic Plants*, Satish Serial Publication, Delhi, 2006.

V.V. Vasiliev.: *Advanced Mechanics of Composite Materials*, Elsevier Publications, Delhi, 2010.

Vishu Arora: *Textile Chemistry*, Abhishek Publications, Delhi, 2010.

# Index